Contents

Experiment Design and Statistical Methods

For behavioural and social research

David R. Boniface
University of Hertfordshire, Hatfield, UK

CHAPMAN & HALL
London · Glasgow · Weinheim · New York · Tokyo · Melbourne · Madras

Published by Chapman & Hall, 2–6 Boundary Row, London SE1 8HN, UK

Chapman & Hall, 2–6 Boundary Row, London SE1 8HN, UK

Blackie Academic & Professional, Wester Cleddens Road, Bishopbriggs, Glasgow G64 2NZ, UK

Chapman & Hall GmbH, Pappelallee 3, 69469 Weinheim, Germany

Chapman & Hall USA, One Penn Plaza, 41st Floor, New York NY 10119, USA

Chapman & Hall Japan, ITP-Japan, Kyowa Building, 3F, 2-2-1 Hirakawacho, Chiyoda-ku, Tokyo 102, Japan

Chapman & Hall Australia, Thomas Nelson Australia, 102 Dodds Street, South Melbourne, Victoria 3205, Australia

Chapman & Hall India, R. Seshadri, 32 Second Main Road, CIT East, Madras 600 035, India

First edition 1995

Typeset in Times 10½/12pt by Interprint Limited, Malta
Printed in Great Britain by Clays Ltd., St. Ives Plc., Bungay, Suffolk.

ISBN 0 412 54230 7

A catalogue record for this book is available from the British Library

Library of Congress Catalog Card Number: 94-70924

∞ Printed on permanent acid-free text paper, manufactured in accordance with ANSI/NISO Z39.48-1992 and ANSI/NISO Z39.48-1984 (Permanence of Paper).

Preface

The subject of the book is in the broad area of statistics. More precisely, it deals with topics of quantitative research methods needed, most commonly, for research with human subjects.

The book focuses on the design of experiments and the analysis of experiments and surveys for quantitative research. It is relevant to small and large scale research both in real-world settings and in laboratories.

The book is intended as a textbook for courses in quantitative research methods and as a self-study and reference book for the postgraduate student or professional researcher in psychology, health or human sciences.

Material is presented at a sufficiently conceptual level to enable the user to be confident in applying the material in a variety of contexts.

The book concentrates on decision-making and understanding rather than on calculation and derivations. It is assumed the user has access to an appropriate computer package such as Minitab, SPSS, SAS, Statview, Super-ANOVA, CSS, BMDP, SYSTAT, Genstat etc.

The main applications of the book are in psychology, education, human, social and life sciences, medicine, and occupational and management research.

This is a second level text. The reader is expected to have previously attended a course in basic statistics or to have read an introductory textbook. This results in the book being more concise than other books in this area.

It introduces the concepts, principles and techniques needed by the empirical researcher or student carrying out a practical project. The exercises which accompany the explanatory material enable the reader to develop competence with the concepts and techniques.

The book deals thoroughly, yet without recourse to mathematics, with several important topics which are usually treated in either a superficial 'cookbook' form or in a heavily mathematical manner. These include:

Repeated measures designs
Unbalanced designs
Non-randomized designs
Model building and partition of variance
Covariate adjustment and multiple regression
Elimination of the effects of nuisance variables
Simplified decision tools for choice of design or analysis

Power and efficiency are treated from a practical point of view showing how they are affected by choice of design, category and continuous covariates and sample size.

A unique extension of the Venn diagram is introduced as an aid to understanding the unbalanced design.

The book is arranged in three parts. Part One reviews the basic concepts of statistics relevant for design and analysis and covers the principles and practice of four basic designs appropriate to research based on experiments. These designs are applicable to a range of situations in which the researcher has a degree of control over the conditions. Analysis of variance, which underpins all these research designs, is developed by an intuitive rather than a mathematical approach.

Part One also includes sections on comparisons and contrasts and on power, sensitivity and sample size and the associated decision-making.

Part Two develops the basic designs discussed in Part One in order that they can be applied to research carried out in field and workplace settings or where the researcher has limited control over the situation.

It includes sections on unbalanced analysis of variance, multiple regression and the elimination of the effects of factors which undermine the validity of research studies.

These techniques include the methods for surveys and comparisons based on non-equivalent groups often required in social or health research or marketing.

Part Three extends the basic designs of Part One to situations where, in research under controlled conditions, more factors are required or the same individuals contribute measurements on more than one occasion. These designs are central to the work of the professional researcher carrying out experiments under controlled conditions in laboratories or community or workplace environments.

There are exercises at the end of each chapter from Chapter 4 onwards. These are carefully matched to each chapter's content. A separate appendix of exercises is located after the final chapter. Many of these further exercises draw on material from several chapters. Worked solutions are provided to many of the exercises.

Acknowledgements are due to members of the Psychology Division at the University of Hertfordshire for several sets of data used as examples.

My thanks also go to the approximately 400 students who, over a number of years, helped me by serving as a sceptical and critical audience for my teaching.

Next, they go to those who provided assistance with the production of the text: the wonderful Margaret Tefft, whose tireless efforts made light of a huge task; Hilary Laurie, who tried to show me how to write about technical ideas for a non-technical audience; Jessica Bennett who tidied up the text; Josie who typed day and night; colleagues Ian Cooper, who helped organize the exercises, Mike Beasley, who read early drafts and gave sound advice; and Michaela Cottee who identified errors in the language and logic of the final draft.

Finally, they go to Pamela Welson who continued to help and believe in me even while the work was going badly.

Statistical Design and Analysis for Basic Experiments

PART 1

Introduction 1

1.1 STRUCTURE AND SCOPE OF PART ONE

1.1.1 Structure

This chapter sets out the framework in which the material of this part of the book is located and identifies the aims of the design of experiments.

Chapter 2 presents examples of each of the four experiment designs dealt with. It includes an introduction to some of the concepts and issues relevant to them.

Chapter 3 presents the concepts of design and analysis for experiments in a degree of detail sufficient for understanding the later material.

Chapters 4–7 each deal in detail with one of the four designs that were introduced in Chapter 2.

Chapter 8 extends the analysis of the designs of Chapters 4–7 to suit them to particular research issues which occur commonly in practice.

Chapter 9 is concerned with the number of individuals to be included in the research and the choice of appropriate design.

1.1.2 Scope

Part One introduces designs, analyses, principles and techniques for comparing alternative conditions in experimental research.

In all experiments dealt with it is assumed that the response of the individuals taking part is measured on a continuous scale. A continuous scale is one in which the numerical values refer to an underflying continuum of amount or quantity. It is further supposed that the measurement scale has the equal value interval property (i.e. one unit has equal value over the whole scale).

The reader is assumed to have completed a basic non-mathematical course in statistical methods and to be familiar with the basic ideas of hypothesis testing, *t*-tests, correlation and regression.

1.2 INFERENCE FOR DESCRIPTIVE AND EXPERIMENTAL RESEARCH

Descriptive research is essentially an exercise in gathering data. The data may be gathered by direct observation, questionnaire or some other method. Some considerable intervention in the lives of individuals may be involved: for example, they may be asked to keep a diary or follow a special diet. Such intervention is made only to provide the conditions under which the observations are to be made; the intervention is not made in order to provide a comparison with the absence of intervention or with some alternative form of intervention.

In descriptive research the design could take one of several forms. It may be a case study; for example, an account of the development of speech in a child with a particular learning difficulty. It may be a study of a sample of individuals; for example, a survey of the extent of examination nerves in a sample of students.

Sometimes research is carried out with very limited aims. A nursing manager may want to carry out a small research project whose end result will be an improved oganization of a hospital ward. In this case there may be no intention to generalize the results of the research to other hospital wards. Very often, however, the researcher wishes to obtain knowledge from the research which can be applied elsewhere. This is true whichever form of descriptive research design is used. In other words, the researcher intends the findings of the particular study to be generalized to other individuals or situations.

Generalizing the results of research can be based on common-sense judgements of the similarity of situations. Such judgements have an important place in scientific work. However, there is also available a formal method for generalizing the findings from descriptive research. This is the method of statistical inference.

Statistical inference uses the mathematics of probability to decide whether the findings of the study are generalizable to the wider population of individuals from which the study sample was drawn. If this inferential form of generalization is to be used, appropriate features need to be designed into the study. The main requirement is that the sample of individuals used in the research be taken randomly from the appropriate population of individuals (see section 3.3) and be of sufficient size.

Descriptive research has an important role in both inferential and non-inferential forms. Its limitation, however, is that it is not capable of establishing that a particular behavioural or environmental factor causes a particular effect or response in the individuals studied.

1.3 WHAT IS EXPERIMENTAL RESEARCH?

Experimental research is characterized by the researcher arranging an intervention in the lives of individuals in order to assess its impact on them. In this text an experiment is understood to be a formally arranged **intervention** which aims to identify **cause–effect** relationships. The interventions are usually referred to as **experimental conditions**. The effects of different interventions are compared. If the interventions are delivered according to proper experimental procedure it may be possible to conclude that the nature of the intervention or condition (the **independent variable** or i.v.) causes an effect in some aspect of the individuals (the **dependent variable** or d.v.).

For example, an experiment could show that the extent of availability of **sample examination papers** (the i.v.) has a causal influence on the amount of **examination nerves** (the d.v.).

Experimental research requires both the proper experimental procedures and the appropriate sampling to ensure that inferential generalization is available. The main requirement for proper experimental procedures is that individuals be randomly allocated to the conditions.

1.4 THEORY TESTING, GENERALIZATION AND COST-EFFECTIVENESS

Behavioural science is concerned with the development of theory about behaviour. Since individuals differ, one from another and one group from another group, theory development in this area clearly faces difficulties that are rarely encountered in the physical sciences. A **theory** is a general explanation of a phenomenon. Thus a theory which applied only to the behaviour of the children in one teacher's infant class would have lower scientific value than a theory which applied to all British infant children.

Experiments test theories. A theory is a general statement. It is in this sense that the results of an experiment are generalizable. Likewise, if the theory is true, then the experiment which tests it must be replicable on other occasions and on other samples of individuals.

Sampling fluctuation is the phenomenon for successive samples to differ from each other even though they are taken from the same population. It is difficult, when carrying out experiments on behaviour, to distinguish generalizable, real phenomena from the effects of sampling fluctuation. This problem is particularly severe if the sample is small.

The size of the sample is the main design feature influencing the ability of the experiment to distinguish a real phenomenon from an effect of sampling fluctuation. If the sample is too small the phenomenon or effect arising from the theory being tested is unlikely to be distinguishable from the effect of sampling fluctuation. This is referred to as the problem of low **power** or low **sensitivity**. Experiments should be conducted on large enough samples of individuals to ensure sufficient power but not so large as to be prohibitively expensive to carry out. (See sections 3.8 and 3.9 for discussions of power and sensitivity.)

Obtaining the correct balance of cost and power is the **cost-effectiveness** aim of the design of experiments.

The other main aim is the **validity** aim. There is discussion of this in sections 3.11 (bias) and 10.3.1 (confounding).

2 Introduction to four basic designs

2.1 SINGLE-FACTOR INDEPENDENT GROUPS DESIGN

The **single-factor independent groups design** refers to an experiment in which members of a sample of individuals are randomly allocated to various conditions. The design is also known as the **between-subjects design**. This name derives from the fact that the comparison between different conditions is a comparison **between** groups of subjects. The purpose of the experiment is to compare the effects of the different conditions on the individuals. An individual's response to a condition is expected to manifest itself through the scores or values of a scale or measure which is known as the **dependent variable**.

Mean scores are obtained under the influence of the conditions and the mean scores of the groups are compared. Differences among the means of the groups are taken as an indication of possible differences among the effects of the conditions.

Random allocation of individuals to conditions is used. This is an intervention in individuals' lives. It is the distinguishing feature of experimental research. It is an essential component of the design if **causal inferences** are required.

The various conditions are assumed to be comparable. All, therefore, may have the same effect. The researcher may hope that the conditions differ, but the possibility that they do not must be tenable. (Otherwise there would be no need for the experiment.)

The set of comparable conditions included in the experiment is known as a **factor**. The conditions that constitute the factor are sometimes known as the **levels** of the factor.

The **effect of the factor** refers to the differences in mean scores of the various groups of individuals influenced by the conditions.

The factor is also referred to as an **independent variable** or i.v. It is a category-type i.v. because the levels of the factor serve to categorize individuals.

For example: it is required to compare the number of words remembered from a list under different time pressure conditions in order to investigate the effect of time pressure on recall for words. The three levels of the factor are:

1. No time instructions given, the subject is asked to read the list at his or her **own speed**.
2. The subject is asked to read the list in **five seconds**.
3. The subject is asked to read the list in **ten seconds**.

The dependent variable is the number of words recalled from the list under test conditions.

Thirty randomly selected individuals (experimental subjects) are allocated at random, ten to each of the three conditions.

Note that **random selection** and **random allocation** of subjects are required to conform to the sampling and proper experimental procedures referred to in sections 1.2 and 1.3. This ensures that inferential generalization is available and that the experiment is capable of identifying a **causal influence** of the independent variable on the dependent variable.

After reading the list each subject's recall is tested and the number of words recalled becomes the score for that subject. The mean scores for the three groups were 5.2, 3.8 and 9.0 words respectively. This result is displayed as a bar chart in Fig. 2.1. The overall mean score in this example is 6.0 words. Hence the apparent effect of the first level of the factor is to lower the scores by 0.8, on average, relative to the overall mean.

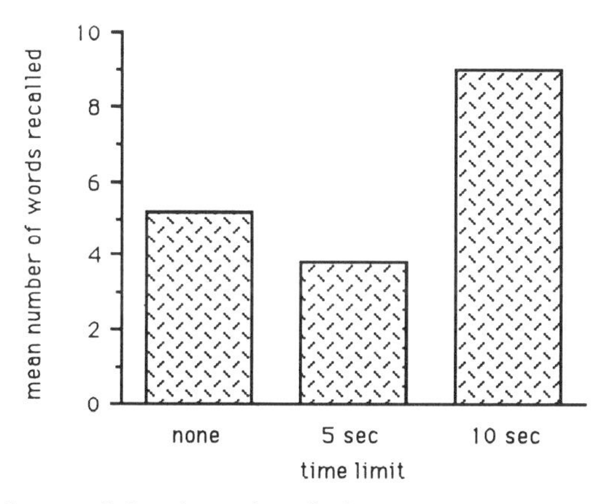

Fig. 2.1 Word list recall for three time limits.

The second and third levels lower and raise the mean score by 2.2 and 3.0 respectively. Hence the apparent effect of the factor can be represented as:

$$(-0.8, \ -2.2, \ +3.0)$$

This bracketed expression is a set of incremental and decremental elements which add to zero and contain the information about the size and direction of the effect of the factor provided by the experiment. (The value of the overall mean itself should not be regarded as an effect of the factor in the sense used here.) Throughout this book the incremental/decremental elements that describe the size of the effect of a factor will be referred to as **deviations**.

The differences among the means of the groups were described as the **apparent** effect of the factor because some differences among the means would be expected even if three identical conditions had been used. This follows from the random allocation of individual subjects to the condition groups. The groups differ because they contain individuals who differ; each individual has a unique score.

In other words, expressed more technically, the chance effects of sampling lead to **sampling fluctuation** among the means of the groups. It is to be understood that the apparent effect of the factor is a combination of the pure effect of the factor and the effect of sampling fluctuation. These two effects can be said to be **confounded**.

The statistics technique known as **analysis of variance** (ANOVA) has been developed to assist the experimenter in deciding whether the differences among mean scores associated with the conditions or groups are due to the effect of sampling fluctuation combined with the effect of the conditions or due to the effect of sampling fluctuation alone. The decision that must be made is whether or not there is any pure effect of the factor (this is the real phenomenon discussed in section 1.4). The making of the decision is discussed further in sections 3.7 and 4.2.

2.2 SINGLE-FACTOR REPEATED MEASURES DESIGN

The **repeated measures design** can sometimes serve as an alternative to the single-factor independent groups design introduced in section 2.1. Instead of allocating subjects at random to different groups so that each group experiences one condition, the subjects are kept in a single group and each subject experiences all the conditions in succession.

Whereas in the single-factor design with independent groups the conditions are compared by making between-group or **between-subject** comparisons, in the repeated measures design the conditions are compared by making comparisons within the one group of subjects, or **within-subjects** comparisons.

For example: an experiment is carried out on the interference between functions in the same or different hemispheres of the brain. Subjects were required to compare mean times for balancing a dowel rod on the left-hand index finger under three conditions: **silent**, **speaking** and **humming**. Four randomly sampled individuals took part in the experiment. The dependent variable is the **balancing time**, which is scored in seconds.

Three measurements of the dependent variable are made on each subject. Each subject's balancing times are set out in Table 2.1. The mean scores under the three conditions were 15.6, 8.1 and 9.6 s respectively. This result is displayed as a bar chart in Fig. 2.2.

Table 2.1 Balancing times under three conditions

		Silent	**Speaking**	**Humming**
Balancing times (seconds)	**Subject 1**	10.2	11.9	7.5
	Subject 2	23.9	5.5	7.0
	Subject 3	17.0	6.0	12.1
	Subject 4	11.3	9.0	11.8
	Means	15.6	8.1	9.6

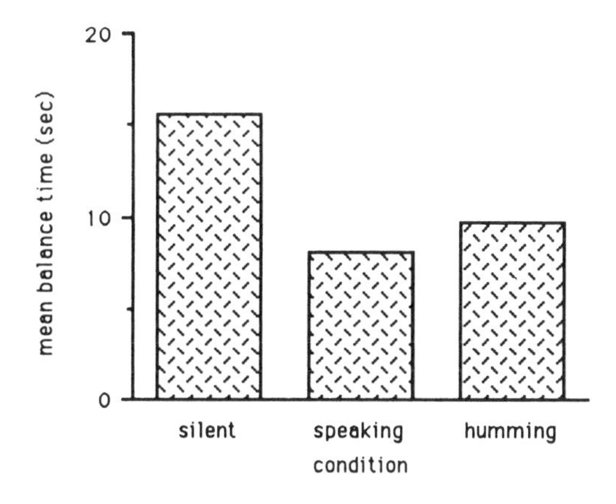

Fig. 2.2 Dowel balancing times for three conditions.

Subtracting the overall mean score from each of the three means gives the apparent effect of the factor, expressed as deviations from the overall mean of 11.1, as:

$$(+4.5, \ -3.0, \ -1.5)$$

As in the case of the independent groups design introduced in section 2.1, this apparent effect of the factor is a combination of the pure effect of the conditions combined with the effect of sampling fluctuation. (Sampling fluctuation in this design refers to randomly selected subjects showing different patterns of response to the conditions. For example, one subject balancing best while humming, another doing best while silent and so on.)

Thus the effect of the factor is confounded with sampling fluctuation in the repeated measures design, as it is in the independent groups design.

The analysis of variance technique is used to assist the experimenter in deciding whether the differences among mean scores of the conditions are due to the effect of sampling fluctuation alone or to sampling fluctuation in combination with a pure conditions effect. This is discussed further in section 4.2.

In general the repeated-measures design is more powerful than the independent groups design, but it is often unusable because of problems arising from the need to obtain scores on the dependent variable several times on each subject. Typical problems are tiredness of subjects, drop-out and practice effects.

However, there is no random allocation of subjects to conditions in this design. This means that differences among the mean scores shown not to be due to sampling fluctuation are not necessarily due to differences among the effects of the conditions. Alternative explanations need to be considered based on considerations of the timing and sequencing of the experiencing of the conditions by each individual. The design can be strengthened by allocation of the conditions in random order to each individual subject.

2.3 TWO-FACTOR DESIGN

2.3.1 Introduction

The two-factor design is an arrangement of conditions which enables the same individuals to serve as subjects simultaneously in the investigation of two distinct factors, each with several levels. This arrangement can only be used if the same dependent variable is used throughout.

Example of a two-factor experiment

An experiment was carried out to examine the effects of type of teaching and type of counselling on children with behaviour and reading problems. A random sample of 40 children from the appropriate population was randomly allocated, 10 to each of four groups. Each group received one of the two conditions from each of Factor 1 and Factor 2:

Factor 1: **Type of counselling**
- level 1: *Individual* for $\frac{1}{4}$ h
- level 2: *In groups* for 1 h

Factor 2: **Type of teaching**
- level 1: *Withdrawal* from normal class
- level 2: *Stay* in normal class

The dependent variable is the improvement in reading score after 15 weeks experience of the allocated conditions. The four groups are displayed with their mean improvement scores as **cells** on the **layout diagram** in Fig. 2.3.

		Factor 2: Type of teaching	
		Withdrawal	**Stay** in class
Factor 1: Type of **counselling**	**Individual**	+1.7	+4.5
	Group	+5.5	+5.6

Fig. 2.3 Layout diagram for two-factor design.

Each subject is measured under the combined influence of two conditions: one which is a level of the first factor and one which is a level of the second factor. For example, the group of subjects represented by the cell in the top right-hand square in Fig. 2.3 experiences the *stay in class* type of teaching and the *individual* type of counselling, and on average the ten children in the group improve their reading score by 4.5 points.

Such a design makes possible the comparison of the two types of teaching for all the subjects regardless of the type of counselling they experienced. This comparison is known as the **main effect** of the factor. This factor is called **type of teaching.** A research question that could be answered by reference to the magnitude of this main effect would be: 'Does the type of teaching influence improvement in reading scores?'.

In numerical terms it can be seen that the mean improvement score for the 20 children experiencing the *withdrawal* from class teaching is $(1.7+5.5)/2=3.6$ and the equivalent value for the 20 *stay in class* children is 5.05. Hence the *stay in class* approach appears to be better. This result is displayed as a bar chart in Fig. 2.4.

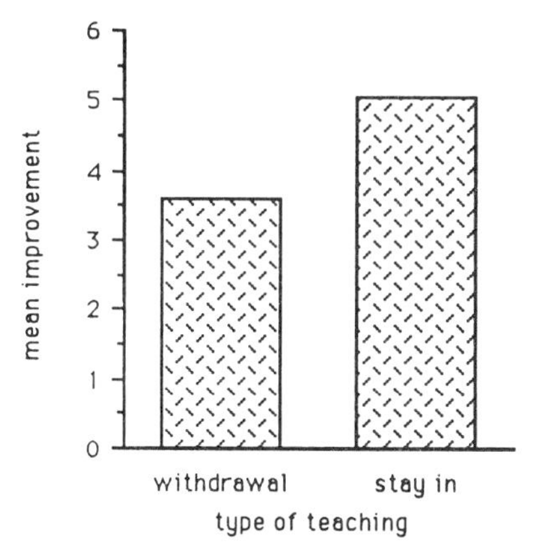

Fig. 2.4 Main effect of type of teaching on reading score improvement.

As for the single-factor designs described in sections 2.1 and 2.2, however, it is possible that differences among the means of the four groups of children are due solely to sampling fluctuation with no contribution from the conditions under which the children are taught. The analysis of variance technique described in section 4.3 estimates the variation due to sampling fluctuation. This makes possible the identification of the portion of the variation among the means that is due to the effect of the conditions.

The comparison of the two types of teaching is also possible, restricted to the subjects who received *individual counselling*. This comparison is known as the **simple effect** of the **type of teaching** under the *individual counselling* condition.

A research question that could be answered by reference to the magnitude of this simple effect would be: 'Does the type of teaching influence improvement in reading scores for pupils receiving individual counselling?'.

The answer is based on the comparison of the values 1.7 and 4.5. Apparently, the type of teaching does affect the improvement in reading scores for the *individual counselling* children. Note, however, that the type of teaching apparently has almost no effect for the *group counselling* children. One simple effect is quite large, the other is almost non-existent. Figures 2.5(a) and 2.5(b) illustrate these two simple effects.

Also available are the main effect and two simple effects of the **type of counselling** factor. Additionally the interaction of the two factors can be investigated.

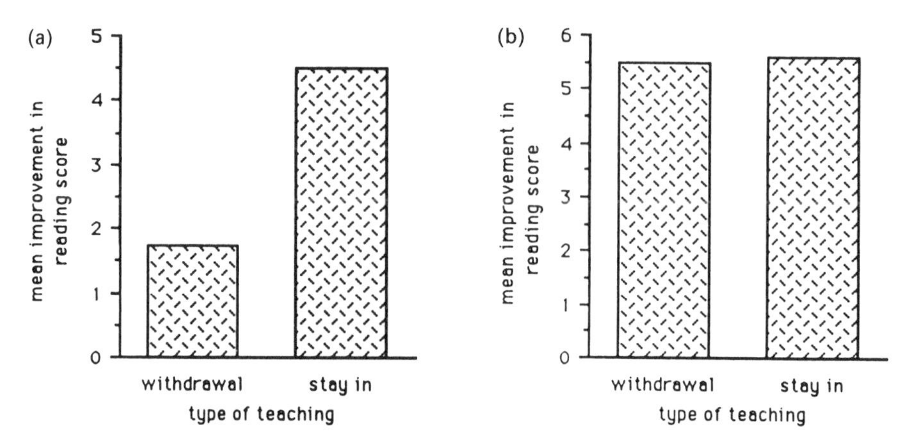

Fig. 2.5 Simple effect of teaching type for (a) individual and (b) group counselling.

Interaction

The interaction is equally the extent to which the two simple effects of **type of teaching** differ from one another and the extent to which the two simple effects of **type of counselling** differ from one another.

A research question that could be answered by reference to the magnitude of the interaction would be: 'Is the benefit of group counselling relative to individual counselling more marked for pupils receiving withdrawal remedial help than for those receiving remedial help staying in their normal class?'.

The answer to this question appears to be 'yes', since for the *withdrawal* children the benefit is $(5.5-1.7)=3.8$ points whereas for the *stay in class* children the benefit is only $(5.6-4.5)=1.1$ points. Figure 2.6 displays this comparison.

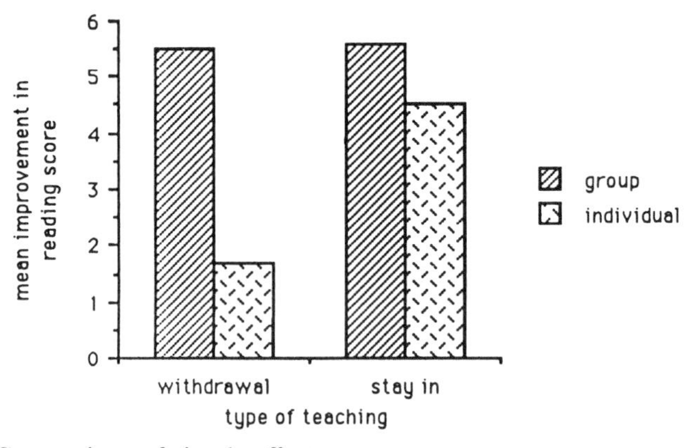

Fig. 2.6 Comparison of simple effects.

See section 6.3 for a discussion of interaction and simple effect.

2.3.2 Randomized block design

This is a special version of the two-factor design in which only one of the factors is the focus of the investigation. The second factor is included to facilitate the study of the first. This second factor is referred to as a **blocking factor** or as a category-type covariate ('category-type' because its levels represent categories to which subjects belong and 'covariate' because its levels correspond to variation in the dependent variable).

The blocking factor has the effect of making the scores of the subjects in any one group or cell more **homogeneous**, which in turn increases the power and sensitivity of the design. There are two types of blocking factor:

1. It may be an **intrinsic factor**, such as the sex of the subjects, in which case the experiment can be viewed as a single-factor design run several times with separate and homogeneous groups of subjects.
2. It may be an **extrinsic factor**, such as day of the week or which of a group of interviewers carried out the interview, in which case the experiment can be viewed as a single-factor design run several times under different conditions.

Figure 2.7 illustrates these two types of blocking factor.

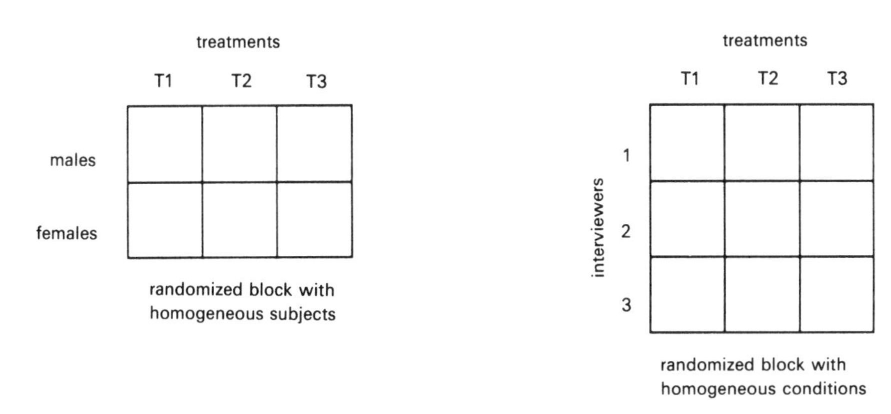

Fig. 2.7 Layout diagrams with different types of blocking factor.

In both cases the same increase in power could have been achieved by either of:

1. Restricting the subjects to a single homogeneous group; for example, males only.
2. Restricting the conditions to greater uniformity; for example, a single day of the week or single interviewer. Such a restriction, however, would have the effect of limiting the generalizability of the findings.

This design is known as the **randomized block design** because subjects are allocated at random to the conditions whilst being organized into several distinct blocks. The advantage of the randomized block design is that it makes possible a more powerful or more sensitive test of a factor without sacrificing generalizability of the findings or economy. See section 3.8 for a discussion of power.

2.3.3 Reasons for using a two-factor design

There are four reasons for using a two-factor design instead of either one or more single-factor designs.

1. *Cost-effectiveness*
   Subjects are simultaneously taking part in two experiments. This is beneficial for cost-effectiveness.
2. *Interaction*
   Additional information is provided on the interaction between the two factors.
3. *Power or sensitivity*
   A second factor may increase the power or sensitivity of the test of the factor being investigated (see sections 3.8 and 3.9 for a discussion of power and sensitivity). This is achieved by introducing a second factor which is known to have an effect on the dependent variable.
4. *Combining single-factor experiments*
   A two-factor design can combine the results of several single-factor experiments into a single analysis. For example, suppose an educational experiment was conducted as a single factor design on successive cohorts of pupils or in several schools and it is required to carry out a single test of the hypothesis that the conditions factor has an overall effect on the scores on the dependent variable. Then it is only necessary to regard the cohorts or schools as the different levels of a blocking factor and the whole as a two-factor design for the desired result to be obtained.

The analysis of variance and test of hypotheses for the two-factor design are discussed in Chapter 6.

2.4 SINGLE-FACTOR INDEPENDENT GROUPS DESIGN WITH USE OF COVARIATE

The randomized block design introduced in section 2.3.2 leads to increased power because the subjects in any one cell are more homogeneous with respect to their scores on the dependent variable. This follows because the blocking factor, a category-type variable (e.g. sex) is related to the scores on the dependent variable.

A similar situation can arise if some continuous-type variable (e.g. IQ) is known to be related to the scores on the dependent variable. Such a variable is called a **concomitant variable** or **covariate**.

The technique of **analysis of covariance** (**ANCOVA**) adjusts the scores on the dependent variable to take account of the values of the covariate by a regression-like technique. This makes the individual subjects taking part in the experiment appear to be more homogeneous. This in turn has the effect of reducing the effect of sampling fluctuation and so increases the power and sensitivity of the design.

This design is very useful provided the cost of obtaining the covariate scores is not too high and the covariate has a **linear** (i.e. straight-line) relationship with the dependent variable.

For example: rats' pulse rates under stress were tested after treatment with either drugs **A** or **B**. Pulse rate was known to depend on the weight of the rat, as shown in Fig. 2.8. (Note that this graph shows the approximate straight-line relationship which is required for the ANCOVA technique.)

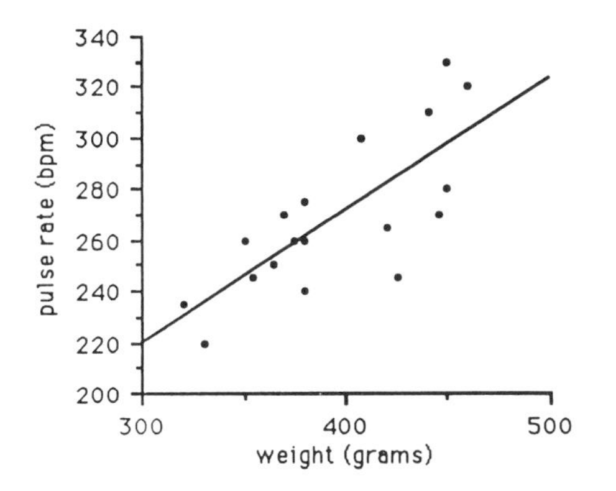

Fig. 2.8 Pulse rate versus weight for rats.

Eighteen randomly selected rats were allocated at random to drug treatment group **A** or **B**. After the experiment the results were as set out in Table 2.2 and displayed in Fig. 2.9.

Table 2.2 Pulse rates and weights of 18 rats

Rat no.	**Drug**	**Pulse**	**Weight**	*Rat no.*	**Drug**	**Pulse**	**Weight**
1	A	330	460	10	B	330	450
2	A	290	450	11	B	310	440
3	A	285	380	12	B	300	408
4	A	280	370	13	B	270	445
5	A	275	420	14	B	260	380
6	A	270	375	15	B	245	425
7	A	270	350	16	B	240	380
8	A	260	365	17	B	235	320
9	A	245	355	18	B	220	330
		Mean 278.3				Mean 267.8	

Parallel straight lines are fitted by regression separately to the A and B plotted data points. The lines are used to adjust the pulse rates in each group to what they would be if the rats had identical weights. The adjusted pulse rates are displayed in Fig. 2.10. Notice how much more homogeneous are the adjusted pulse rates as compared to the unadjusted pulse rates.

The result of the experiment is to find that drug **A** leads to a mean pulse rate of 278.3, whereas drug **B** leads to a mean pulse rate of 267.8.

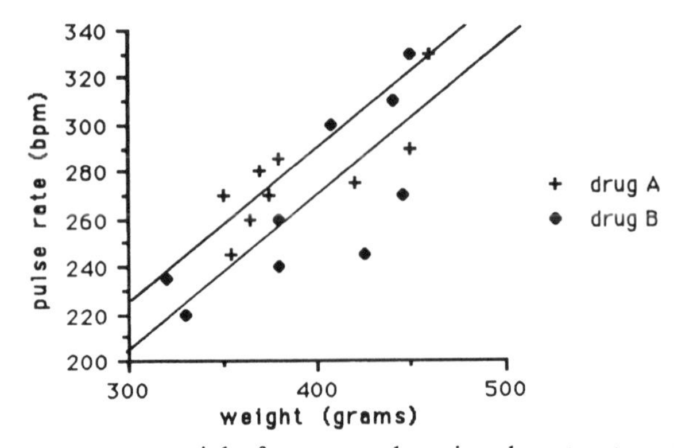

Fig. 2.9 Pulse rate versus weight for rats undergoing drug treatment.

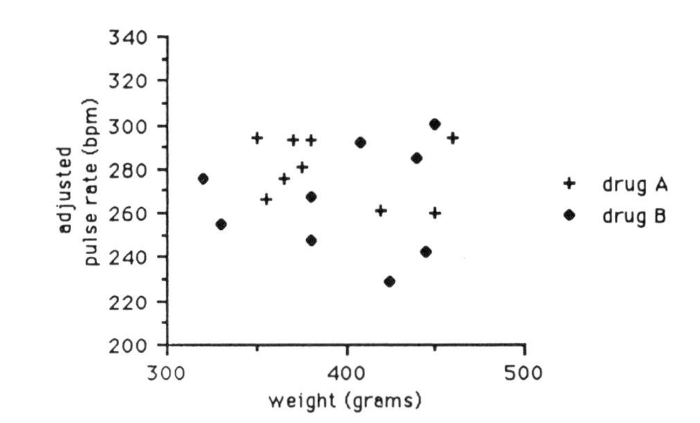

Fig. 2.10 Pulse rates adjusted for weights for drug-treated rats.

The analysis of variance for the single-factor design with covariate (ANCOVA) is discussed further in Chapter 7.

Overview of concepts and techniques

3

3.1 VARIANCE

Variance is a measure of spread or scatter in a group of scores. Variance is based on the sizes of the deviations from the mean of each of the scores in the group. Hence a group of identical scores has a variance of zero. More precisely, **variance** is the **mean of the squared deviations**.

For example, consider the balancing times of the four individuals in the **silent** condition in the example in section 2.2.

Scores (in seconds)	10.2	23.9	17.0	11.3	(mean = 15.6)
Deviations (score − mean)	−5.4	8.3	1.4	−4.3	

Note that the deviations add to zero. This follows from the nature of the mean.

The variance is the mean of the squared deviations:

$$\text{Variance} = \frac{(-5.4)^2 + (8.3)^2 + (1.4)^2 + (-4.3)^2}{4}$$

$$= \frac{29.16 + 68.89 + 1.96 + 18.49}{4} = 29.625$$

The **sum of squared deviations** is often known as *SS* or just **sum of squares**. It is sometimes referred to as the **corrected** sum of squares to distinguish it from the sum of squares of the raw scores.

Estimating variance

When the purpose of the variance calculation is to estimate the variance of a population from a small sample the formula is modified. The sum of squared deviations, instead of being divided by n, the number of deviations, is divided by $(n-1)$, the number of independent deviations. The general term for the number of independent deviations is **degrees of freedom**. In the above example, it is evident that not all four deviations are independent. This follows since they are known to add to zero. If it were known that the first three were −5.4, 8.3 and 1.4, the fourth one would have to be −4.3. So only $(n-1)$ or three are **free**. In other words the degrees of freedom are 3. **Degrees of freedom** is often abbreviated to *df*.

When a variance is being estimated the formula is often seen in the following form:

$$\text{variance estimate} = \frac{SS}{df}$$

This is sometimes called a **mean square** and abbreviated to *MS*. The square of the Greek letter sigma is usually used to stand for a value of a population variance. It is written σ^2. Commonly s^2 is used for the value of an estimate of a population variance based on sample data.

When analysing data from experiments, variances of means are of interest. Variances of means are related to variances of scores by a simple relationship. This is discussed in the next section.

3.2 VARIANCE OF MEANS

When a population of individuals is sampled several times the result is a number of equivalent but different groups of individuals. If each individual contributes a score then there is a mean score for each group. These group means will, in general, differ. Variance is used to measure the amount of difference or spread among the group means.

If the scores in the sampled population have a variance represented by σ^2 then the means of samples of n individuals (i.e. n subjects per group) will have a variance equal to

$$\frac{\sigma^2}{n}$$

This is called the **variance of means** and is represented by the symbol σ^2_{means}.

Most analysis of variance (ANOVA) is discussed in terms of estimates of the variance of scores obtained from variances of means. In other words, the reverse form of the above formula is used:

$$\sigma^2 = n(\text{variance of means})$$

The sum of squared deviations part of this is calculated as:

$$SS = n(\text{sum of squared deviations among means})$$

The multiplier n in the above formula often causes puzzlement. The logic for it, however, is straightforward. It is that the variance of individual scores is being analysed. The n is a weight used to scale up the estimate from an estimate of the variance of means to an estimate of the variance of individual scores.

Example of SS calculation

Take the example data from the single-factor independent groups design in section 2.1. There are 10 subjects per group and three groups, whose means are 5.2, 3.8 and 9.0. The overall mean is 6.0.

The deviations among the means are found by subtracting the **mean of means**, which is 6.0, from each of the three means to get:

$-0{\cdot}8 \quad -2.2 \quad 3.0$

These are squared for insertion into the above formula:

$$SS = 10(0{\cdot}8^2 + 2.2^2 + 3.0^2) = 144.8$$

It will be seen that all mean squares encountered in analysis of variance are estimates of variances of individual scores in the sampled population. Not all are equally good estimates, however.

3.3 RANDOM SAMPLING AND RANDOMIZATION

Random sampling

In so far as research aims to discover or establish truths that are in some sense general truths, two conditions must prevail. Firstly, there must be a **defined population** of individuals to which the truths are to apply. The size of this population and its durability over time influences the scientific value of the truths. Secondly, the individuals investigated, whether by experiment or survey, must be **randomly sampled** from this population.

Random sampling requires that each individual member of the population has the same chance of being selected for inclusion in the sample. Most behaviour research is carried out on subjects easily accessible to the researcher. These subjects form a **sub-population**. They are not a proper random sample from the population to which the findings are to be generalized. This does not mean that any attempt at random selection should be abandoned. Rather, the experimenter should select randomly from the sub-population and accompany the write-up of the research with a discussion of possible differences between the intended target population and the sub-population.

For example, suppose the intended target population is the nation's students, and students taking lunch in a college refectory form the available sub-population; then the researcher should devise a procedure for random sampling of diners from the refectory. Failure to do this introduces bias of unknown degree into the findings.

Randomization

It is desirable that the results of an experiment be attributable to no other causes than the random effects of sampling fluctuation or to the effects of the factors designed into the experiment or to the combined effect of both these. In order to ensure that no other factor, known or unknown, could be having an influence on the dependent variable, randomization must be used in the conduct of the experiment. (Such a factor is known as a **confounding factor**.)

This means that individual subjects must be assigned at random to the different conditions and that random selection of materials, stimuli, interviewers, times of day, rooms etc. must be used whenever these are not prescribed by the design of the experiment or by logistical constraints.

3.4 CONFIDENCE INTERVALS

A mean score is often obtained from a sample of individuals and used as an estimate of the mean score in the wider population from which the sample was taken. An indication of how good an estimate is provided by the sample mean can be provided by the confidence interval.

The confidence interval is a range of values above and below the sample mean so constructed as to have a 95% or 99% chance, or probability, of containing the **true** or **population** value of the mean. In other words the confidence interval is a guide to how close the estimate is likely to be to the true value. The true value can be conceptualized as the value approached by the mean as the sample size increases to include the entire population.

In the context of experiments of the types described in sections 2.1–2.4, approximate confidence intervals can be constructed for means obtained under experimental conditions in the following way.

Consider the word recall scores from the example in section 2.1. The mean number of words recalled by the 10 individuals in the first condition is 5.2. Suppose the analysis of variance has obtained a mean square for within-groups (see section 4.1) whose value is represented by MS. Then the 95% confidence interval is

$$\text{Confidence interval} = 5.2 \pm (1.96) \times \left(\frac{MS}{n}\right)^{1/2} \tag{3.1}$$

In this formula, n takes the value 10, the number of recall scores that have been averaged to obtain the mean value 5.2. The plus provides the upper limit above 5.2 and the minus the lower limit below 5.2. The sample mean itself, 5.2, is the best estimate of the population or true value.

Identifying the appropriate mean square from the analysis of variance needs some skill; however, a rule of thumb is to take the MS with the largest df (degrees of freedom). It may be called MS within-groups, MS error or MS between subjects.

It is often useful to mark the upper and lower 95% confidence limits on each bar on a bar chart of means. Some computer programs will do this.

The 99% approximate confidence interval is obtained by substituting 2.58 for 1.96 in the above formula. (*Note*: ± 1.96 and ± 2.58 are the values of the standardized normal distribution which enclose 95% and 99% of the population.)

3.5 SAMPLING FLUCTUATION AND SAMPLING ERROR

Since every individual has unique properties and abilities, each will return a unique score on any test or measurement. It therefore follows that the mean scores of the groups to which individuals are randomly allocated will differ from one another in a random manner. This is what is meant by **sampling fluctuation**. It is also called **sampling error**.

Sampling fluctuation refers to the changes in value of the mean as repeated random samples are drawn from the same population. These sample means can be considered as a collection of estimates of the true value. Each of them deviates from the true value to a greater or lesser extent. These deviations are errors of estimation, hence the name 'sampling error'.

3.6 STATISTICAL SIGNIFICANCE

If, in an experiment based on a random sample of individuals, differences among means are large enough to be judged to be the result of real differences among the conditions, then these differences are said to be **statistically significant**.

Equivalently, statistical significance is said to be present if the differences found in a sample are large enough to be generalized to the population with confidence.

If a difference in means has been declared to be significant a decision has been made. Whether the decision has been made that a difference in means is or is not significant there is some probability that the decision is in error. The **level of significance** is the probability that a difference in means has been erroneously declared to be significant. Typical values for significance levels are 0.05 and 0.01 (corresponding to 5% and 1% chance of error). Another name for significance level is ***p*-value**.

3.7 FORMULATING DECISION-MAKING AS A TEST OF HYPOTHESES

The experiment used as an example in section 2.1 has as its aim the making of a decision as to whether any differences among the mean scores of the various groups of individuals are due (at least in part) to the effects of the different amounts of time pressure they have experienced. In other words, the aim is to determine whether there is any effect of the time pressure on the recall.

Commonly, researchers ask, 'Is the effect of the independent variable on the dependent variable statistically significant?'.

More concisely, the aim can be stated as being to decide whether the time pressure (the i.v.) is having any effect (on the d.v.). This is a 'yes' or 'no' issue which is often formulated in terms of two hypotheses, one of which proposes that the i.v. is not having an affect (called the **null hypothesis**, H_0) and the other which proposes that the i.v. is having an effect (called the **alternative hypothesis**, H_1):

H_0: time pressure does not have an effect on recall
H_1: time pressure has an effect on recall

or, more generally:

H_0: the i.v. does not have an effect on the d.v.
H_1: the i.v. has an effect on the d.v.

or, in other words and omitting mention of the d.v.:

H_0: the factor does not have an effect
H_1: the factor has an effect

or, equivalently:

H_0: the conditions have indentical effects
H_1: the conditions have different effects

Note that H_0, the null hypothesis, must refer to the absence of effect of the i.v. on the d.v., whereas the alternative hypothesis must refer to the opposite situation. It is supposed that H_0 is taken to be true until the results of an experiment lead to a decision to reject H_0 in favour of H_1.

Two further formulations are commonly used, each useful for its reference to underlying concepts:

H_0: $\mu_1 = \mu_2 = \mu_3 = \ldots$ etc.
H_1: not H_0

where μ_1 is the mean score in the population after exposure to condition 1, and so on. Sampling fluctuation cannot affect the values of μ_1, μ_2 etc. because they are the mean values that would be obtained if the entire population was taking part in the experiment. When the entire population is included there is no sampling fluctuation.

The formulation of H_0 and H_1 in terms of means μ_1, μ_2, etc. being either identical or not identical is equivalent to saying that the conditions either have or do not have identical effects.

Taking this one step further, stating that the population values of the means do not differ is equivalent to stating that they have a zero variance. Hence, if σ^2_{means} is the variance of μ_1, μ_2, μ_3, etc., the equivalent formulation is:

H_0: $\sigma^2_{\text{means}} = 0$
H_1: $\sigma^2_{\text{means}} \neq 0$

(where $\neq$ means 'is not equal to').

All of the above six equivalent formulations are regularly used by practitioners and appear in standard textbooks and journal articles. None is more correct than any other.

At the conclusion of the analysis the decision is reported in terms of rejection or non-rejection of H_0 at a conventional level of significance or accompanied by the computer-calculated p-value. The conventional levels of significance are 0.05, 0.01 and 0.001 (i.e. 5%, 1% and 0.1%).

Examples of reporting the decision

The decision must be accompanied by a statement of the significance level or *p*-value, as in these examples:

H_0 was rejected at the 0.05 significance level.
H_0 was not rejected at the 0.01 level of significance.
H_0 was rejected at the 5% level.
H_0 was not rejected; $p=0.831$.
H_0 was rejected; $p=0.003$.
H_0 was rejected; $p<0.01$.
H_0 was not rejected; $p>0.05$.

The meaning of $p=0.831$ is that the differences among the means are of such a size that deciding to reject H_0 would be wrong 83.1 times in 100. Likewise, $p=0.003$ means that the differences among the means are of such a size that deciding to reject H_0 would be wrong 0.3 times in 100. (See section 3.6 on statistical significance.) It follows from the *p*-values in these two examples that H_0 should not be rejected in the first but should be rejected in the second.

The meaning of $p<0.01$ is that the decision is to reject H_0 at the 0.01 level of significance. The meaning of $p>0.05$ is that the decision is to not reject H_0 at the 0.05 level of significance.

Note that the result is never reported in terms of acceptance of H_0 or rejection of H_1.

3.8 POWER

Experiments pose the problem of distinguishing real effects of the conditions from the effects of sampling fluctuation (see section 1.4).

The design of experiments aims to maximize the effect of the conditions on the dependent variable relative to the effect of sampling fluctuation. The more this is achieved, the more powerful is the experiment.

The analysis of experimental data by analysis of variance provides information in a form that enables the researcher to decide whether or not there is an effect of the treatment factor or conditions. This is the same as deciding that the differences among the means under different conditions are statistically significant. As discussed in section 3.6, it is possible that the wrong decision is made. Power has a direct bearing on the probability of deciding that there is no effect of the conditions when in fact there is an effect. This is called the **type II error**. It can be contrasted with the **type I error** – deciding that there is an effect of the conditions when there is none.

Type II error is likely when the sampling fluctuation is large. This can occur when the individual subjects taking part in the experiment are very heterogeneous. It can also occur when the sample size is small, since in small samples the naturally occurring differences between the subjects may be so large as to obscure the effect of the conditions.

Type II error is also more likely when the conditions being investigated have little effect on the individual scores on the dependent variable. This can be because the true effects of the conditions are small or because of measurement error in the dependent variable.

Formally, power is defined as the probability that there will not be a type II error, i.e the probability of correctly deciding that there is an effect of the conditions. If power is too low it is not worth carrying out the experiment. Conventionally, designers of experiments seek levels of power in excess of 0.7.

3.9 SENSITIVITY

Power can be increased indefinitely by increasing the number of individual subjects taking part in the experiment. It is useful to look for ways of increasing power by changes to the design of the experiment rather than by increasing the number of subjects.

Sensitivity is more convenient than power for comparing designs of alternative experiments which investigate the same conditions. Sensitivity is defined as the number of subjects experiencing each experimental condition divided by the variance of scores in the sample. It is the same expression as that of which the square root was taken in equation (3.1), except that it is the other way up, namely:

$$\text{sensitivity} = \frac{n}{MS}$$

Here n is the number of individual subjects experiencing each condition and MS is the mean square estimate of variance of individual scores.

Sensitivity, then, increases when n increases and decreases when MS increases. Note that MS is a measure of sampling fluctuation. It is often known as **mean square error** or **mean square residual**.

The link with the confidence interval formula referred to above means that as sensitivity reduces, the confidence interval widens, indicating that estimates have larger margins of error. Thus sensitivity relates in a direct way to precision of estimation.

There is an example of the calculation of sensitivity in Chapter 9.

3.10 EFFICIENCY

Since the sensitivity of any design can be increased indefinitely by increasing the number of subjects, the experimenter usually has to consider sensitivity relative to the cost of running the experiment. To serve this end, **efficiency** is defined as follows:

$$\text{efficiency} = \frac{\text{sensitivity}}{\text{cost}}$$

Costs are usually measured in terms of time and can be expected to include the following:

1. Cost of finding subjects
2. Cost of taking subjects through the conditions
3. Cost of setting up conditions
4. Cost of obtaining covariate scores (if available)

The comparison of alternative designs can be carried out in terms of their **relative efficiency** or R.E.:

$$\text{relative efficiency} = \frac{\text{efficiency of design version 1}}{\text{efficiency of design version 2}}$$

The use of relative efficiency depends on the assumption that an alternative design is preferred provided it leads to an increase in sensitivity which is proportionately greater than the increase in costs.

There is an example of the calculation of relative efficiency in Chapter 9.

3.11 BIAS

Bias is systematic error as opposed to sampling error. Sampling error is the tendency of a sample not to mirror the population from which it is drawn because of the chance effects of random sampling. The effects of sampling error diminish towards zero as the size of the sample is increased. Bias is a form of error which does not diminish as the sample size increases.

In a cross-reference to psychometrics, bias is to validity what sampling error is to reliability. Bias will arise if the technique for drawing a random sample is faulty, or if there is a mismatch between the data and the assumptions of the model on which the statistical analysis technique is based. Sometimes it is possible to make an adjustment to correct for bias. One technique for this is dealt with in Part Two of this book.

3.12 LOGISTICAL CONSTRAINTS

There are always limitations on the amount of environmental and economic resources, such as rooms, equipment and time, and on the properties of experimental subjects, such as motivation, availability and resistance to tiredness.

The experiment must be designed to fit within these constraints. Decisions to this end resemble decisions aimed at pursuing any project in the real world and, like them, become easier with experience.

4 Single-factor independent groups design

4.1 INTRODUCTION

A more complete and detailed account of the design introduced in section 2.1 now follows. The design was illustrated in section 2.1 by an investigation of the effect of time pressure on recall of words read from a list. The aim of the experiment was to enable a decision to be made as to whether **time pressure**, the independent variable, caused changes in the number of **words recalled**, the dependent variable.

Section 4.2 sets out the principles of analysis of variance (ANOVA) for the single-factor design. It contains an account of the logic of the process for making a decision about the possible existence of an effect of **time pressure** on recall.

In section 4.3 the principles presented in section 4.2 are illustrated by their application to a new example of the single-factor design. The example is concerned with the eating behaviour of gerbils.

Section 4.4 explains the ANOVA summary table.

Section 4.5 presents convenient formulae for hand calculation of the analysis. This section may be ignored by those readers preferring to use an appropriate computer system.

Finally, in section 4.6 the assumptions which underlie the analysis of the single-factor design are identified and discussed. It is shown that a precise mathematical model is assumed which relates the independent variable to the dependent variable.

4.2 THE PRINCIPLES OF THE ANALYSIS OF VARIANCE

When the null hypothesis is true the various groups of subjects can be seen as random samples from the same population. In the example referred to previously this is equivalent to the different amounts of **time pressure** having identical effects on the number of **words recalled**.

Suppose that the population has mean score μ and variance σ^2. (σ^2 is the between-subjects variance.) Suppose also that the random samples each contain n subjects (the sample size of each group is n). This is represented as a diagram in Fig. 4.1. In this situation the fundamental property of sampling distributions states that if the means are themselves regarded as

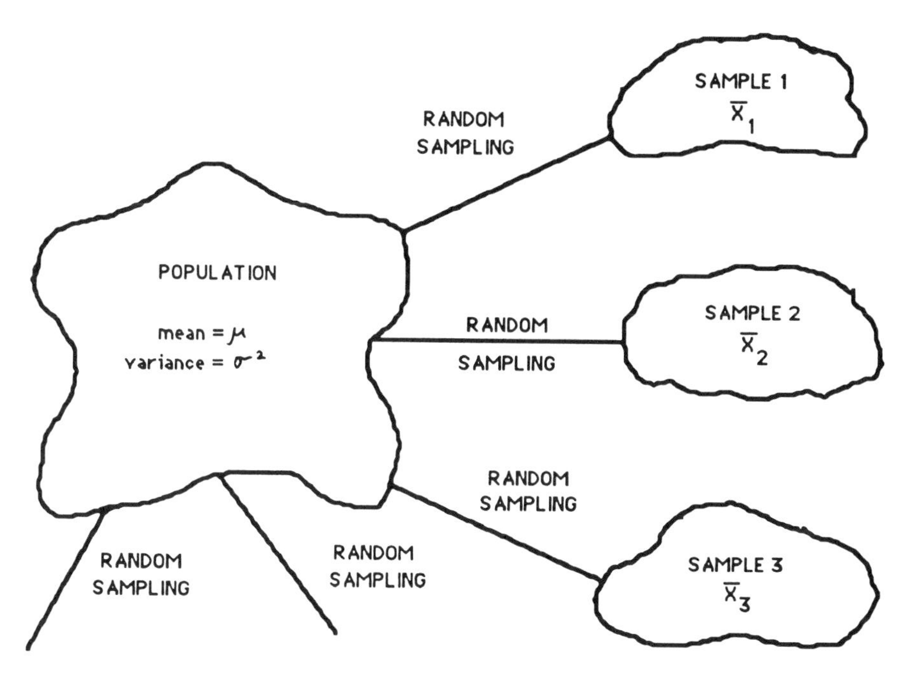

Fig. 4.1 Random samples from a population.

a group of scores they form a random sample from a population of such means whose mean is μ and whose variance (the variance of means discussed in section 3.2) is:

$$\sigma^2_{\text{means}} = \frac{\sigma^2}{n}$$

The significance test of the analysis of variance is based on the comparison of the estimate of σ^2 obtained from n times the variance of means, as discussed in section 3.2, with the estimate of σ^2 obtained from the individual scores within each group. This latter estimate is formed by combining the separate estimates of σ^2 from each group. Combining separate estimates is called **pooling.**

The estimate based on the scores within the groups is not affected by the differences among the means of the groups and so is independent of the truth or falsity of H_0.

The other estimate, however, is affected by the truth or falsity of H_0, for if H_0 is false the group means will exhibit an additional degree of scatter or variation due to the differential effects of the conditions. It will be an overestimation of the between-subjects variance. This leads to the result:

Estimate of variance based on differences among group means	=	Estimate of variance based on scores within-groups

if H_0 is true, or

Estimate of variance based on differences among group means	>	Estimate of variance based on scores within-groups

if H_0 is false.

The ratio of these two **variance estimates** is called F:

$$F = \frac{\text{variance estimated between group means}}{\text{variance estimated from scores within-groups}}$$

F is the statistic which is calculated as part of the ANOVA technique. If H_0 is true, F is expected to have the value 1; if H_0 is false, F is expected to exceed 1.

It is not expected that the value of F from any single realization of the experiment will be exactly 1, even if H_0 is true. F is subject to sampling fluctuation. Mathematical probability theory has made possible the calculation of values of F (known as 'critical values') which are exceeded with probability 0.05 and 0.01 when H_0 is true.

The **critical value of *F*** is the upper limit which will be exceeded in only 5% or 1% of realizations of the experiment with H_0 true. If F exceeds the critical value the decision is made to reject H_0 in favour of H_1. The critical values for 5% and 1% significance levels of the sampling distribution of F are set out in tables in Appendix F.2. The critical value for 5% is displayed on a diagram of the sampling distribution of F in Fig. 4.2. (The critical value of F depends on degrees of freedom – see sections 3.1 and 4.3.1.)

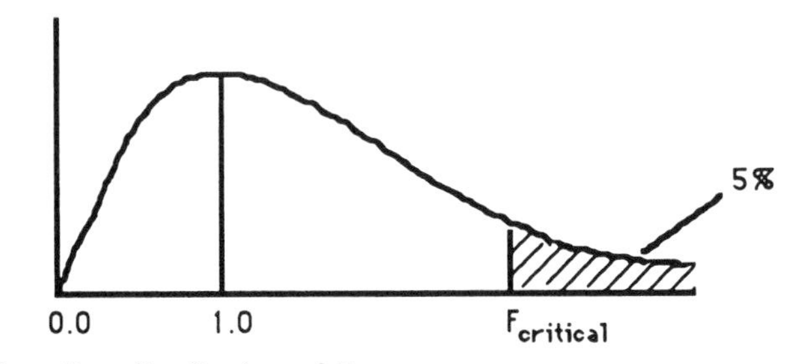

Fig. 4.2 Sampling distribution of F.

4.3 ANALYSIS OF VARIANCE AND SIGNIFICANCE TEST

4.3.1 Numerical example

An experiment aimed to investigate the effect of interrupting gerbils' feeds on their decisions to return to the same feeding site. Thus the conditions factor was the **degree of interruption**, with the three groups each being treated to one of three different degrees of interruption (*none*, *partial* or *complete*). The response or dependent variable was the percentage of times each gerbil subsequently returned (**returns**) to the original feeding site in the next 24 hours.

Twenty-four gerbils, randomly selected from a defined population, were randomly allocated to the three conditions. Thus there were three groups of 8 gerbils (k, the number of groups = 3; n, the number of gerbils per group = 8). The null and alternative hypotheses, expressed in words are:

H_0: the **degree of interruption** does not have an effect on **returns**
H_1: the **degree of interruption** has an effect on **returns**

The results were as set out in Table 4.1. The mean percentage of times the gerbils returned to the original feeding site according to condition groups are set out in Table 4.2 and displayed as a bar chart in Fig. 4.3.

Table 4.1 Percentage returns by feeding condition for 24 gerbils

Factor:	**Degree of interruption**		
Levels:	**None**	**Partial**	**Complete**
	63	61	38
	53	59	44
	53	55	47
	38	75	38
	9	75	59
	47	63	41
	28	53	75
	19	44	34

Table 4.2 Mean percentage returns

Degree of interruption	*Mean*
None	38.75
Partial	60.63
Complete	47.00
Overall mean:	48.79

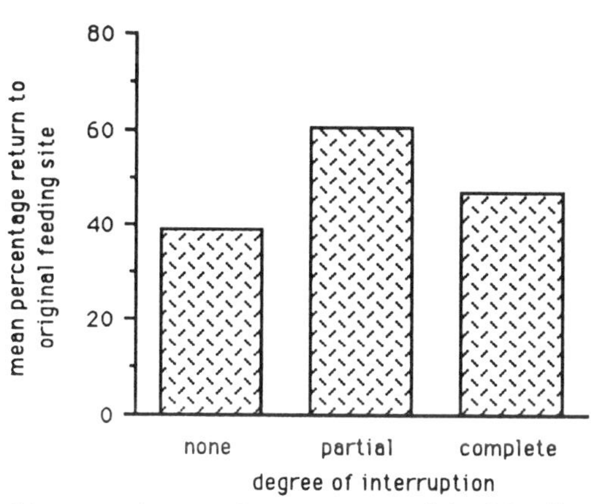

Fig. 4.3 Effect of interruption on the return to original feeding site.

4.3.2 Algebraic formulations of variance estimates

The between-groups variance – symbolic form

One of the two variance estimates referred to above is that obtained from the means of the k groups. If the group means are represented by $\bar{X}_1$, $\bar{X}_2$, $\bar{X}_3$, ..., $\bar{X}_k$, and $\bar{X}$ represents their overall mean (mean of means) then the deviation of the jth mean from the overall mean is $(\bar{X}_j - \bar{X})$. The sum of squares of all such deviations is set out as

$$SS = \Sigma(\bar{X}_j - \bar{X})^2$$

summed over all groups. It is an SS which, when divided by the appropriate degrees of freedom, df, estimates σ^2/n as discussed in sections 3.2 and 4.2. When multiplied by n, supposing there are n scores per group, it provides an SS which when divided by the appropriate degrees of freedom estimates σ^2. It has the form:

$$SS = n\Sigma(\bar{X}_j - \bar{X})^2$$

This is the ***SS* between-groups**, which can be written SS_{between}. It has $k-1$ degrees of freedom. Hence the between-groups variance estimate (known as MS_{between}) is

$$MS_{\text{between}} = \frac{n\Sigma(\bar{X}_j - \bar{X})^2}{k-1}$$

Numerical illustration
Following the calculations of section 3.2 as formulated above, this gives, for the gerbil experiment:

$$\begin{aligned} SS &= 8[(38.75 - 48.79)^2 + (60.63 - 48.79)^2 + (47.00 - 48.79)^2] \\ &= 8(244.19) \\ &= 1953 \end{aligned}$$

This is the SS between-groups. When divided by the degrees of freedom, $k-1$, in this case 2, it gives 976 as the estimate of the variance of individual scores known as the Mean Square between-groups.

The within-groups variance – symbolic form

Also referred to in section 4.2 is the pooled within-groups variance estimate. Suppose the scores in the jth group are represented by X_{1j}, X_{2j}, X_{3j}, ..., X_{nj}, so that the typical score is X_{ij}, that is, the score of the ith gerbil in the jth group. This means that, in the gerbil example, X_{11}, is 63, X_{41} is 38, X_{12} is 61 and X_{83} is 34. Suppose, as before, that $\bar{X}_j$ is the mean of the scores in the jth group, so that $\bar{X}_1$ is 38.75, etc.

Then a typical deviation of an individual score from the appropriate group mean is $(X_{ij} - \bar{X}_j)$ and the SS pooled from all such deviations is

$$SS = \Sigma\Sigma(X_{ij} - \bar{X}_j)^2$$

summed over all scores i and groups j. This is the ***SS* within-groups**, which can be written SS_{within}. It has $k(n-1)$ degrees of freedom. It follows that the within-groups variance (known as MS_{within}) is estimated by

$$MS_{\text{within}} = \frac{\Sigma\Sigma(X_{ij} - \bar{X}_j)^2}{k(n-1)}$$

Numerical illustration

The within-groups *SS* is obtained by summing the squares of the deviations of the scores each from their own group means. The deviations from the group mean of the first two scores are: (63 − 38.75) and (53 − 38.75). There are three groups of eight gerbils, each contributing one deviation. The sum of squares of all 24 such deviations is 4545.

$$\begin{aligned} SS &= [(63-38.75)^2 + (53-38.75)^2 + \cdots + (34-47.00)^2] \\ &= 4545 \end{aligned}$$

Only the first two and the last terms are shown.

This is the *SS* within-groups. When divided by the degrees of freedom, $k(n-1)$, in this case 21, it gives 216 as the estimate of the variance of individual scores known as the mean square within-groups.

4.4 THE SUMMARY TABLE AND THE DECOMPOSITION OF THE TOTAL *SS*

4.4.1 Symbolic form

The sum of squared deviations, which is known as *SS* for short, as described above, is a very convenient measure of variation on which to base an analysis of the results of an experiment. This is because of the existence of the decomposition of *SS*.

Before the decomposition of *SS* can be fully appreciated, one further *SS* formulation is required. It is the *SS* obtained by supposing that all scores from the *k* groups belong to a single group containing *nk* scores. The *SS* obtained from these *nk* scores is called SS_{total}.

The analysis is based on the algebraic relationship between SS_{total}, SS_{between} and SS_{within}. The relationship amounts to a decomposition of the total *SS* into two components as follows:

$$SS_{\text{total}} = SS_{\text{between}} + SS_{\text{within}}$$

Thus when variation is measured in terms of *SS*, a decomposition of the total variation is provided into a component due to differences between the means of the groups and a component due to differences between the scores within the groups.

The ANOVA **summary table** provides a standard way of displaying this decomposition of total variation together with the variance estimates and the *F*-statistic described in section 4.2. The variance estimates are referred to as mean squares in the table (abbreviated to *MS*). There is an equivalent decomposition of the total degrees of freedom into the sum of the between- and within-groups *df*.

Each of the SS, when divided by its degrees of freedom, provides a **variance estimate** or **mean square**.

If H_0 is true, all three are estimates of σ^2, the population value of the between-subjects variance of the scores.

If H_0 is not true, the within-groups mean square continues to estimate σ^2, but the between-groups mean square estimates

$$\sigma^2 + n(\sigma^2_{\text{means}})$$

where σ^2_{means} is the variance of the population values of the group means described in sections 3.2 and 4.2.

Thus the ratio of the between- to the within-groups mean squares, known as the F ratio, is dependent on the size of σ^2_{means}. The larger is σ^2_{means}, the larger is F.

When H_0 is true the means of the groups in the population are identical and σ^2_{means} is zero and F is expected to equal 1.0.

4.4.2 Numerical illustration of summary table

A numerical example taken from the gerbil experiment will assist the explanation. First, the SS_{total} is calculated. In practice this is rarely needed except as a check on other calculations.

The deviations from the overall mean of the first two scores are $(63-48.79)$ and $(53-48.79)$, and the sum of squares of all 24 such deviations is 6498, with 23 degrees of freedom (the degrees of freedom for SS_{total} are always one fewer than the number of measurements in the experiment).

Thus 6498 is the total SS obtained by disregarding the group membership of the gerbils and supposing them to belong to a single group and to have experienced identical conditions.

It is now possible to show the decomposition of SS for this example:

$$SS_{\text{total}} = SS_{\text{between}} + SS_{\text{within}}$$
$$6498 = 1953 + 4545$$

There is an equivalent decomposition of dfs. Note that the formulae for degrees of freedom are presented in section 4.5. The complete analysis is displayed in a summary table as in Table 4.3. The **critical value of F** on 2 and 21 degrees of freedom at the 5% level of significance is 3.47. It is shown on a graph of the distribution of F in Fig. 4.4. The 2 and 21 degrees of freedom are referred to as the numerator and denominator degrees of freedom since they are the dfs of the MSs which form the numerator and

Table 4.3 Summary table for analysis of variance for the gerbil experiment

Source of variation	*SS*	*df*	*Mean square*	*F*
Between-groups	1953	2	976	4.51
Within-groups	4545	21	216	
Total	6498	23		

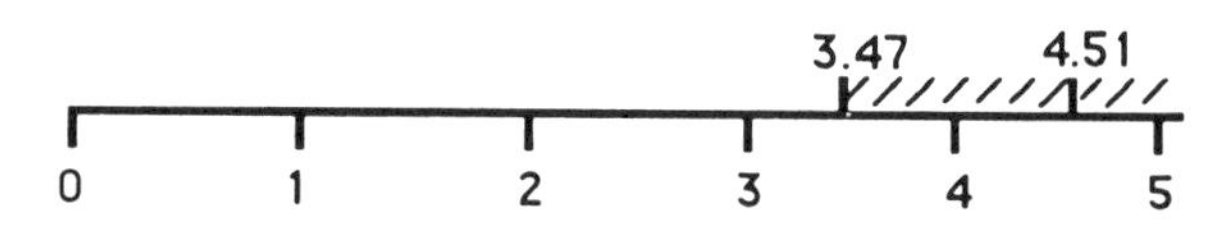

Fig. 4.4 Line graph for F.

denominator of the calculation of F. When looking up the critical F in the tables, the numerator and denominator *df*s determine the column and row respectively. (Note that it is safe to assume that the denominator *df* is larger than the numerator *df* for this design.)

Therefore, since the value of F found from analysis of the results of the experiment exceeds the critical F value, the decision is made to reject H_0 at the 0.05 level of significance.

Note that the variance estimate obtained between-groups is 4.5 times the size of the within-groups estimate. This intuitively suggests that the variation among the group means is far greater than is likely to be obtained due to the chance effects of the random sampling. The intuition is confirmed by the significance test. The variance of the population means can be estimated using the formula in section 4.4.1: $(976-216)=760$ estimates n times σ^2_{means}. Hence

$$\begin{aligned}\sigma^2_{\text{means}} &= 760/8\\ &= 95\end{aligned}$$

This value, 95, is a measure of the size of the effect of the factor on the dependent variable. A more usual way of describing the size of the effect is set out in section 4.4.3.

4.4.3 Size of effect

The size of the effect of a factor can be described in terms of the variance of the means in the various conditions as above. In this approach $\sigma^2_{\text{means}}=95$.

An easier approach is to express the *SS* due to the conditions factor as a proportion of the SS_{total}. In the current example this leads to the fraction 1953/6498, indicating that 30% of the variation in gerbils' scores is explained by the differences among conditions.

4.5 COMPUTATIONAL FORMULAE FOR DEGREES OF FREEDOM AND *SS*s

The method of calculation used in the above example illustrates the principles of ANOVA, but is subject to a serious build-up of rounding errors and is slow. A better method involves calculating the sums and sums of squares of the scores in each group. That is, where n is the number of individuals in each group and N is nk, the total number of individuals across all groups, we calculate:

$$SS_{\text{between}} = \frac{(\Sigma X_{i1})^2}{n} + \frac{(\Sigma X_{i2})^2}{n} + \frac{(\Sigma X_{i3})^2}{n} - \frac{(\Sigma X_{i1} + \Sigma X_{i2} + \Sigma X_{i3})^2}{N}$$

$$=\frac{(310)^2}{8}+\frac{(485)^2}{8}+\frac{(376)^2}{8}-\frac{(310+485+376)^2}{24}$$

$$=1953$$

as before, and

$$SS_{\text{total}}=\Sigma\Sigma X_{ij}^2-\frac{(\Sigma X_{i1}+\Sigma X_{i2}+\Sigma X_{i3})^2}{N}$$
$$=63\,633-57\,135$$
$$=6498$$

as before. SS_{within} is just $6498-1953=4545$.

Degrees of freedom

For the total *SS*, the *df* is always one less the total number of measurements of the dependent variable: $df=N-1$.

For the between-groups *SS* the *df* is always one less than the number of groups: $df=k-1$.

For the within-groups *SS* the *df* is always (the number of groups) × (one less than the number of measurements per group): $df=k(n-1)$.

4.6 UNDERLYING MODEL AND ASSUMPTIONS FOR TESTS OF SIGNIFICANCE

Since the effect of a factor has been defined in terms of differences between means there is an implicit assumption that an additive effects model is being used. That is:

score = overall mean + conditions effect

The conditions must be conceptualized as having uniform incremental or decremental effects on the subjects' scores. In the numerical example of section 4.3.1 this model is estimated as:

Expected score for a randomly sampled subject = overall mean + conditions effect

$$=48.79 \qquad +\begin{cases}-10.04\\+11.84\\-\ \ 1.79\end{cases}$$

The conditions effect is −10.04 for the first condition, +11.84 for the second and so on. These values are the deviations of the group means from the overall mean.

When differences are the focus of interest then sums of squared deviations are an obvious statistic for the measurement of the size of a collection of differences. This follows since differences are deviations. Variances are a measure of the average size of the squares of deviations: hence **mean square**.

The various concepts that make up analysis of variance are interdependent and follow from a particular conceptualization of the conditions effect.

The test of significance depends for its validity on the sampling design of the experiment, on the additive nature of the conditions effect and on the scores having a normal distribution in the population. A condition which affected individuals differently depending on their score levels before experiencing the condition is not permissible. (An example of such a condition would be a form of remedial reading tuition which only benefits the weakest readers.)

The combined effect of these three requirements is that the scores in each of the conditions groups is assumed to be a random sample from a population of scores with a normal distribution with a variance identical to that in all other conditions groups. It is only in the means of their parent populations that the groups can differ and these differences must be due only to the effects of the conditions.

4.7 CONCEPT LINKAGE FOR ANALYSIS OF VARIANCE

Figure 4.5 represents the links between the concepts used to relate the differences between individuals and between groups to variances and the test of significance.

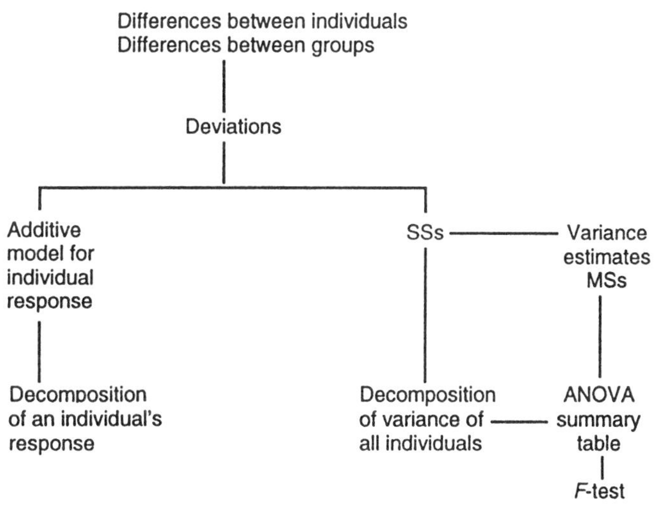

Fig. 4.5 Concept linkage diagram for analysis of variance.

The differences are reformed into deviations from an overall mean (section 3.2). In one development these are used in the additive model (section 4.6). This shows how an individual subject's response (measured as the dependent variable) is represented as the sum of components due to the effects of the conditions and an overall mean.

In a parallel development the deviations are reformed into *SS*s, their weighted sums of squares, and are used in the decomposition of variance (sections 4.3.2 and 4.4).

The *SS*s are also used to obtain the variance estimates, which are known as mean squares, or *MS*s. The *MS*s lead directly to the *F*-test of significance for the differences between means of groups in the experiment conditions (section 4.2).

4.8 EXERCISES

4.1 The result of running a single-factor between-subjects experiment with three conditions and 10 subjects per group was expressed as deviations from the overall mean thus:

effect of conditions = (0.75, −1.9, 1.15)

(a) If the overall mean score on the dependent variable was 7.88, use the model on which the ANOVA is based in the form (section 4.6):

expected score = overall mean + conditions effect

to calculate the score predicted by the model for an individual in the third condition group.

(b) Calculate SS_{between} by using the following formula from section 4.3.2:

$$SS_{\text{between}} = \text{(No. of subjects per group) (sum of squared deviations of group means from overall mean)}$$
$$= n\Sigma(\bar{X}_j - \bar{X})^2$$

where n is the number of subjects per group, $\bar{X}_j$ is mean of group j and $\bar{X}$ is the overall mean.

(c) Sketch the bar chart of mean scores of the three groups.

4.2 Take the data for the gerbil example in Table 4.1 and carry out an analysis using an appropriate computer package. Obtain the following:

(a) The mean scores in the three conditions.
(b) A graphical display of the means in (a) as a bar chart or equivalent.
(c) An analysis of variance summary table with *p*-value or some equivalent indication of significance.

Compare your results with those obtained in section 4.4. Are you clear what the decision is about H_0? Reject or not?

4.3 Rats were randomly allocated to one of the following three conditions:

no experience of novel food
smell of novel food
contact with demonstrator rat used to novel food

The dependent variable was the amount of the novel food eaten in grammes.

There were twenty rats in each group. The mean scores in the respective groups were:

0.255
0.216
0.365

The SSs were:

Between: 0.2389
Within: 4.3403

(a) Express the results as a bar chart of group means.
(b) Obtain the size of the conditions effect as a percentage of the total SS.
(c) Using the rules for degrees of freedom (section 4.5):

$$df_{\text{between}} = k - 1$$
$$df_{\text{within}} = k(n - 1)$$

where k is the number of groups and n is the number of subjects per group, construct the complete ANOVA table, calculate F and complete the test of significance.
(d) Set out the two estimates of individual error variance, σ^2, one derived from differences among group means, the other from differences among individual scores within groups (sections 4.2 and 4.3.2). Comment on their relative sizes in relation to the result in (c).

5 Single-factor repeated measures design

5.1 INTRODUCTION

A more complete and detailed account of the design introduced in section 2.2 now follows.

The design was illustrated in section 2.2 by an experiment on the interference between functions in the same or different hemispheres of the brain. Four randomly sampled individuals took part. The dependent variable, on which all four subjects were measured three times, was the **length of time** they could balance a dowel rod on the index finger of the left hand. Measurements were made, in succession, while subjects were **silent**, **speaking** or **humming**. The scores are set out in Table 2.1.

The repeated measures design differs from the independent groups design in that all subjects are measured under all conditions instead of being allocated into separate groups, each to be measured under only one condition.

Section 5.2 deals with the influences on the score obtained by a particular individual subject under a particular condition. Sampling fluctuation is shown to influence the results in a more complicated way in this design.

Section 5.3 sets out the principles of analysis of variance for the single-factor repeated measures design. Much the same principles apply as in the independent groups design.

The principles presented in section 5.3 are illustrated in section 5.4 by their application to the analysis of the example referred to above.

Section 5.5 presents convenient formulae for hand calculation of the analysis. This section may be ignored by those readers preferring to use an appropriate computer system.

Finally, in section 5.6 the assumptions which underlie the analysis of this design are identified and discussed. It is shown that a precise mathematical model is assumed which relates the independent variable to the dependent variable.

5.2 VARIATION PRESENT IN THE REPEATED MEASURES DESIGN

Figure 5.1 represents, using hypothetical values, the sampling and measurement aspects of the dowel balancing experiment. A, B and C are three

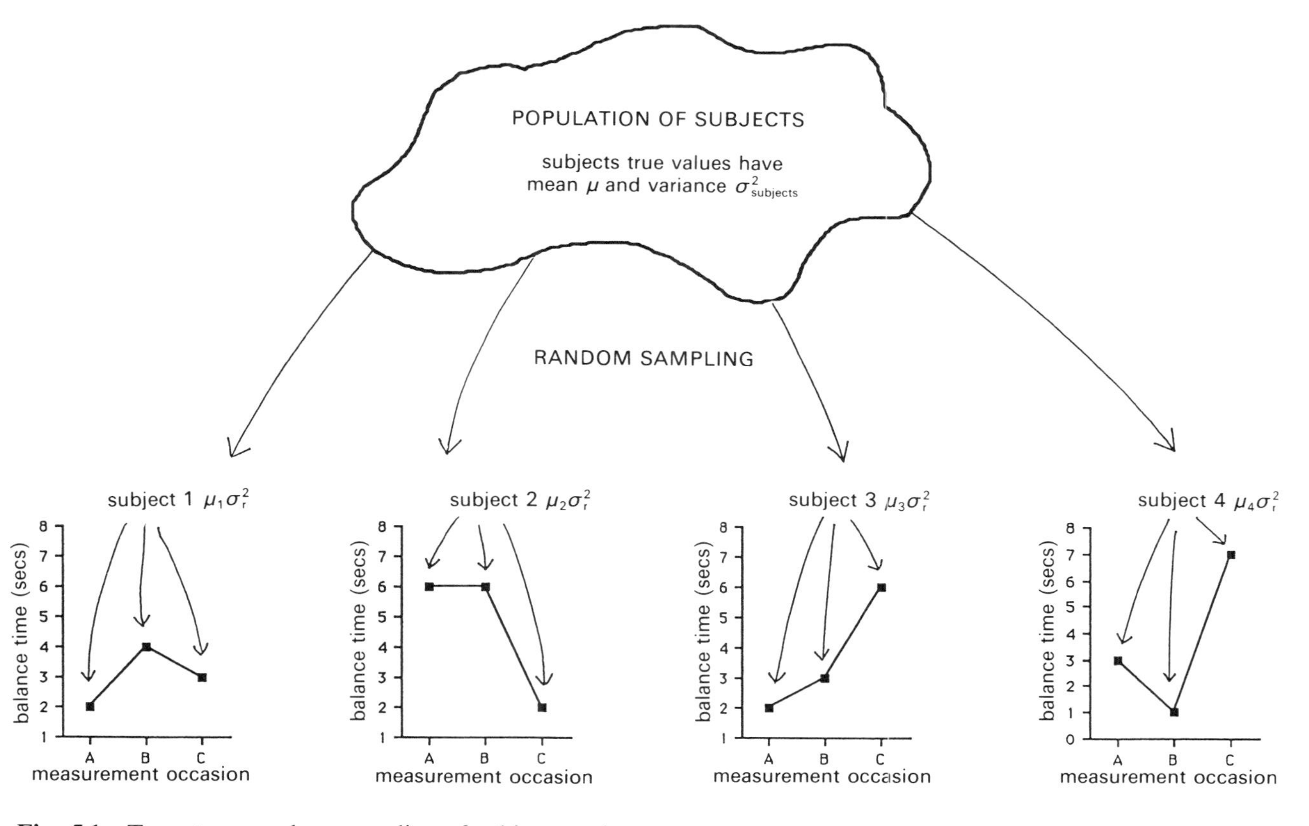

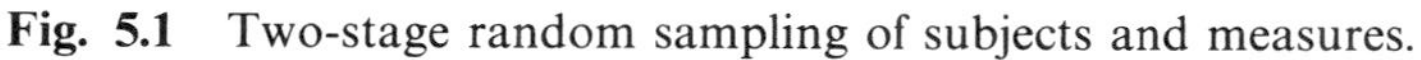

Fig. 5.1 Two-stage random sampling of subjects and measures.

occasions, randomly selected, at which subjects were tested in dowel balancing. It is supposed that the three conditons, **silent**, **speaking** and **humming**, have identical effects on **balancing times** and so can be arbitrarily represented by A, B and C. In other words, H_0 is supposed true in Fig. 5.1

There are two stages of sampling. Each creates a source of sampling fluctuation which has consequences for the analysis and interpretation of the results of the experiment.

The first stage of sampling (illustrated in Fig. 5.1) is the random selection of individual subjects from the defined population. For the particular dependent variable used in the experiment, **balancing times**, the mean score in the population is represented by μ. The differences among individuals are represented in the population by the variance $\sigma^2_{\text{subjects}}$. This variance influences the range of balancing times likely to be obtained in the experiment. It is referred to as **subjects variance** or as **between-subjects variance**.

Four subjects are selected. Each has, thanks to inherited and acquired characteristics, a unique balancing ability and corresponding true balancing time. This true balancing time for each subject is represented in the figure by μ_1 for the first subject, μ_2 for the second and so on.

The second stage of sampling is the obtaining of the three measurements for each subject. Since H_0 is supposed true, the conditions are having no effect. Therefore the three measurements (A, B and C) should be seen as a random sample of three from the supposedly infinite available population of occasions on which measurements could be provided by each subject.

In Fig. 5.1 the values of three hypothetical measures for each subject are shown plotted on line graphs on which the balancing time is represented by the vertical axis.

Since H_0 is supposed true for the purposes of this discussion, the three measurements are obtained under identical conditions. Each of them is an estimate of the true balancing time for the subject. They differ because of the error of measurement and instability or unreliability of the human subject.

The variation shown by successive measurements of **balancing time** on an individual subject is referred to as **reliability variance**. It is represented by $\sigma^2_{\text{reliability}}$ or σ^2_{r}.

In the event that H_0 is not true, the differences between the conditions under which the three measurements are obtained act as another influence on the values of the balancing times.

5.3 THE PRINCIPLES OF THE ANALYSIS OF VARIANCE

The same general idea applies here as in section 4.2 for the independent groups design. That is, two independent sets of deviations estimate an underlying source of variation, and one only of the estimates, the **between-conditions estimate**, is inflated if H_0 is not true. The underlying source of variation is the reliability variance introduced in section 5.2.

The reliability variance can be estimated from the amount of variation shown by individual subjects from measurement to measurement not otherwise accounted for. This estimate is not affected by any effect of the conditions if an arithmetic adjustment is carried out before the reliability variance is estimated. The adjustment removes any effect of the conditions.

The reliability variance can also be estimated from the **between-conditions variance**. This is the variance among the mean scores in the various conditions. n times the between-conditions variance estimates the reliability variance (n is the number of subjects in the group). This estimate is inflated by any effects the conditions might be having and is directly comparable to the between-groups variance in the independent groups design.

As in the independent groups design, comparison of the estimates of the reliability variance from these two sources serves as the basis of the test of significance of H_0. It follows that:

Estimate of variance based on variation among condition means	=	Estimate of variance based on variation within-subjects and within-conditions

if H_0 is true, or

Estimate of variance based on variation among condition means	>	Estimate of variance based on variation within-subjects and within-conditions

if H_0 is false.

The ratio of these two variances is called F and leads to the same test of significance as that discussed in section 4.2:

$$F = \frac{\text{variance estimated between condition means}}{\text{variance estimated within-subjects and within-conditions}}$$

F is the statistic which is calculated as part of the ANOVA technique. If H_0 is true, F is expected to have the value 1; if H_0 is false, F is expected to exceed 1.

It is not expected that the value of F from any single realization of the experiment will be exactly 1, even if H_0 is true. F is subject to sampling fluctuation. Further discussion of this is found in section 4.2.

5.4 ANALYSIS OF VARIANCE AND SIGNIFICANCE TEST

5.4.1 Numerical example

The numerical example is based on the experiment described in section 2.2. The aim is to test the hypothesis that the three conditions (**silence**, **speaking** and **humming**), have identical effects on the length of time a subject balances a dowel rod on the index finger of the left hand.

This is the null hypothesis H_0. It is tested against the alternative hypothesis H_1 which states the opposite, namely that there are differences among the three conditions in their effects on the balancing times.

To simplify the presentation of the calculations, fictitious data has been substituted for the real data shown in section 2.2: see Table 5.1.

Table 5.1 Balancing times under three conditions

		Silent	Speaking	Humming	Means
Balancing times (seconds)	**Subject 1**	8	2	2	4
	Subject 2	11	8	11	10
	Subject 3	9	6	3	6
	Subject 4	12	8	4	8
	Means	10	6	5	7

5.4.2 Reliability variance estimated within-subjects

The variation among the three scores obtained for a particular subject is caused by imperfect reliability augmented by any effects of the conditions.

For example, the values 8, 2 and 2 obtained by the first subject reflect imperfect reliability of that subject together with error of a random nature inherent to the measuring process. Any differences between the effects of the three conditions is also reflected in the values 8, 2 and 2.

The effect of the conditions is estimated by the deviations among the condition means 10, 6 and 5. This is expressed as the **deviations** of the means from the overall mean, 7:

$$\text{Effect of conditions} = \{(10-7), (6-7), (5-7)\}$$
$$= \{3, -1, -2\}$$

The effect of conditions is removed from the **balancing times** in Table 5.1 by literally subtracting 3 from all times in the **silent** condition, -1 from all times in the **speaking** condition and -2 from all times in the **humming** condition. The result is set out in Table 5.2. Note that the mean **balancing times** for each subject have not been affected by the removal of the estimated conditions effect. By contrast, the condition means have changed to identical values, 7, to reflect the absence of conditions effect in the data.

Table 5.2 Balancing times after removal of effects of conditions

		Silent	Speaking	Humming	Means
Balancing times (seconds)	**Subject 1**	5	3	4	4
	Subject 2	8	9	13	10
	Subject 3	6	7	5	6
	Subject 4	9	9	6	8
	Means	7	7	7	7

The remaining variation among the times for each subject represents reliability variation alone.

Consider the first subject. The variations among 5, 3 and 4 expressed as deviations from the subject's mean are

$$\text{Reliability deviations for subject } 1 = (1, -1, 0)$$

The complete set of deviations for all four subjects is set out in Table 5.3. The size of these deviations is a measure of the amount of unreliability in the measurement of **balancing times**.

They relate to the $SS_{\text{reliability}}$ in the obvious way:

$$SS_{\text{reliability}} = \{(1)^2 + (-1)^2 + \cdots + (1)^2 + (-2)^2\}$$
$$= 24$$

Table 5.3 Deviations within subject and within conditions

Subject 1	1	−1	0
Subject 2	−2	−1	3
Subject 3	0	1	−1
Subject 4	1	1	−2

Noting that every row and column must add to zero it is evident that only six of the deviations in Table 5.3 can be independently determined. The values of the remaining six are determined by the row and column totals. This means that there are six degrees of freedom. The general rule is:

$$df_{\text{reliability}} = (k-1)(n-1)$$

Note that k is the number of conditions and n is the number of subjects. Hence, for the balancing time example,

$$df = (3-1)(4-1)$$
$$= 6$$

The same value was obtained previously by reasoning from basic principles.

As seen in section 3.1, the estimate of a variance is obtained as the result of dividing the SS by the appropriate df. This gives $24/6 = 4$ as the mean square, which estimates the reliability variance and which cannot be affected by any effect of the conditions.

5.4.3 Reliability variance estimated between conditions

This estimate is obtained from the differences among the mean scores in the various conditions. Thus in this example it is based on the amount of difference among the means 10, 6 and 5.

It is equivalent to the between-groups variance estimate for the independent groups design described in section 4.2 and is calculated by the formula set out in that section or, equivalently, in section 3.2.

The deviations of the means from the overall mean are $(10-7)$, $(6-7)$ and $(5-7)$, that is, 3, -1 and -2.

The SS between-conditions is obtained as n times the sum of the squares of these deviations, where n is the number of subjects:

$$SS = 4\{(3)^2 + (-1)^2 + (-2)^2\}$$
$$= 4\{14\}$$
$$= 56$$

The between-conditions variance estimate has $k-1$ degrees of freedom where k is the number of conditions; in this example $k=3$, $df=2$.

As seen in section 3.1 the estimate of a variance is obtained as the result of dividing the *SS* by the appropriate *df*. This gives $56/2=28$ as the mean square, which estimates the reliability variance augmented by any effects of the conditions which may exist. If H_0 is true and the conditions have identical effects this *MS* estimates the reliability variance. Hence the same logic applies as in the independent groups design.

The standard *F*-test of significance may now be carried out:

$$F=\frac{MS_{\text{between-conditions}}}{MS_{\text{reliability}}}=\frac{28}{4}=7$$

This must be compared with the critical value of *F* based on 2 and 6 degrees of freedom. The critical values of *F* are 5.14 and 10.9 at the 5% and 1% significance levels, respectively. Since 7 exceeds 5.14, the decision is to reject H_0 at the 5% level. The conclusion follows that the conditions do affect balancing times.

5.4.4 The subjects-by-conditions interaction

The conventional approach to analysis of this design refers to reliability variance as subjects-by-conditions interaction. The identical numerical result is obtained. However, since the subjects-by-conditions interaction cannot be estimated separately from measurement error unless each subject is measured at least twice in each condition it is not a particularly helpful approach in this design.

5.4.5 The summary table and size of effect

The summary table

The complete decomposition of the variation in the scores is usually presented in an ANOVA summary table. For the numerical example above this takes the form of Table 5.4. Of the sums of squares in this table, only the between-subjects one was not calculated in sections 5.4.2 and 5.4.3. It is obtained as in section 3.2 as the sum of squares of the subjects-effect deviations multiplied by k, the number of conditions. The multiplying factor k arises because each subject-effect deviation can be seen as the mean of three deviations arising each under one of the three conditions.

Table 5.4 Summary table for analysis of variance

Source of variation	*SS*	*df*	*MS*	*F*
Within-subjects:				
Conditions	56	2	28	7.0
Reliability	24	6	4	
Between-subjects:	60	3	20	
Total	140	11		

The mean scores obtained by the four subjects are 4, 10, 6 and 8. Expressed as deviations (subject-effect deviations) these are $(4-7)$, $(10-7)$, $(6-7)$ and $(8-7)$, that is, -3, 3, -1 and 1. These are squared, added and multiplied by k to give the SS, which estimates the variation due to subjects' differences, represented by $\sigma^2_{\text{subjects}}$ in Fig. 5.1:

$$\begin{aligned} SS_{\text{subjects}} &= 3[(-3)^2+(3)^2+(-1)^2+(1)^2] \\ &= 60 \end{aligned}$$

The degrees of freedom for subjects is $(n-1)$, in this case 3. Hence the estimate of $\sigma^2_{\text{subjects}}$ is $60/3=20$.

Note that the ANOVA summary table is in two distinct sections, the within-subjects part containing estimates of the reliability variance and the between-subjects part containing an estimate of the subjects variance.

The size of effect

The size of the effect of the **conditions** factor can be reported as the proportion it explains of the total within-subjects SS. In the example above the value is

$$\frac{56}{56+24}=70\%$$

The rationale for expressing the SS for **conditions** as a proportion of the total SS_{within}, rather than as a proportion of the overall total SS, is that **conditions** contributes only to the within part of the overall total SS.

5.5 COMPUTATIONAL FORMULAE FOR SS AND DEGREES OF FREEDOM

If hand calculation is required, the fastest formulae are as follows:

$$SS_{\text{conditions}} = \frac{(\text{Cond.1 total})^2}{n} + \frac{(\text{Cond.2 total})^2}{n} + \cdots + \frac{(\text{Cond.}k \text{ total})^2}{n} - \frac{(\text{Overall total})^2}{nk}$$

$$SS_{\text{subjects}} = \frac{(\text{Subj.1 total})^2}{k} + \frac{(\text{Subj.2 total})^2}{k} + \cdots + \frac{(\text{Subj.}n \text{ total})^2}{k} - \frac{(\text{Overall total})^2}{nk}$$

$$SS_{\text{total}} = \Sigma\Sigma X^2 - \frac{(\text{Overall total})^2}{nk}$$

where k is the number of conditions, n is the number of subjects, Cond. 1 total is the total of scores, in condition 1, Subj. 1 total is the total of scores for subject 1 etc., and $\Sigma\Sigma X^2$ is the sum of squares of every measurement over all conditions and subjects.

The $SS_{\text{reliability}}$ is found by subtraction.

The degrees of freedom follow the usual rules:

1. For SS_{total} the df is one less than the total number of measurements: $df=(nk-1)$
2. For $SS_{\text{conditions}}$ the df is one less than the number of levels of the conditions factor: $df=(k-1)$.
3. For SS_{subjects} the df is one less than the number of subjects: $df=(n-1)$.
4. Finally, for $SS_{\text{reliability}}$ the df is (df for subjects)$\times$(df for conditions)$=(n-1)(k-1)$.

5.6 UNDERLYING MODEL AND ASSUMPTIONS FOR TESTS OF SIGNIFICANCE

Underlying model

As for the independent groups ANOVA, the conceptualization of an effect in terms of deviations implies an underlying additive model. This is set out as:

expected score = overall mean + subject effect + conditions effect

For the numerical example above the model appears as:

Expected score for a randomly sampled subject = overall mean + subject effect + conditions effect

$$= 7 \quad + \begin{Bmatrix} -3 \\ 3 \\ -1 \\ 1 \end{Bmatrix} \quad + \begin{Bmatrix} 3 \\ -1 \\ -2 \end{Bmatrix}$$

The three terms in the above model, when appropriately added, give the value of the score expected supposing there to be no error.

Assumptions for validity of the F-test

1. The subjects are supposed randomly sampled from a defined population.
2. The reliability errors (which are the set of deviations illustrated in Table 5.3) are supposed randomly sampled from a population which has a normal distribution, mean zero and variance $\sigma^2_{\text{reliability}}$.
3. It is required that in the population the scores on the dependent variable follow a normal distribution with variance $\sigma^2_{\text{subjects}}$.
4. Additivity is assumed. See section 4.6 for discussion of additivity.

5.7 EXERCISES

5.1 Five randomly sampled migraine sufferers took part in an investigation of two rival medications. The dependent variable was the number of hours of migraine experienced in one week. Four observations were obtained on each subject – two baseline non-medication weeks followed by one week on medication 1 and one on medication 2.

The results were as follows:

Subject	**Conditions**				**Subject** means	**Conditions** means	Overall mean
	Base		**Medication**				
	1	2	1	2			
1	21	22	8	6	14.25	22.60	15.45
2	20	19	10	4	13.25	22.60	
3	17	15	5	4	10.25	9.80	
4	25	30	13	12	20.00	6.80	
5	30	27	13	8	19.50		

(a) Display the **conditions** effect as a bar chart of mean scores. Which is the more effective medication?

(b) Display the **conditions** effects (as raw scores) unique to each subject superimposed on a single graph. Which **subject** appears to have benefited least from treatment?

(c) Obtain the conditions effect as deviations from the overall mean and hence use the formula:

$$SS_{\text{conditions}} = (\text{no. of measurements at each } \textbf{cond.})(t_1^2 + t_2^2 + t_3^2 + \cdots)$$

(where t_1, t_2, etc. are the deviations) to obtain the SS for **conditions**. (This formula is set out at the between-conditions variance estimate in section 5.4.3.)

(d) The three following tables are the migraine data modified by removing one of: the **subjects** effect; the **conditions** effect; both the **subjects** and **conditions** effects (but not necessarily in that order). Identify which is which. (Hint: obtain the **subject** means and the **conditions** means for each table; refer to section 5.4.2.)

Table 1

15.05	16.05	14.85	15.85
15.05	14.05	17.85	14.85
15.05	13.05	15.85	17.85
13.30	18.30	14.10	16.10
18.80	15.80	14.60	12.60

Table 2

13.85	14.85	13.65	14.65
12.85	11.85	15.65	12.65
9.85	7.85	10.65	12.65
17.85	22.85	18.65	20.65
22.85	19.85	18.65	16.65

Table 3

22.20	23.20	9.20	7.20
22.20	21.20	12.20	6.20
22.20	20.20	10.20	9.20
20.45	25.45	8.45	7.45
25.95	22.95	8.95	3.95

(e) When the **subjects** effect, **conditions** effect and overall mean are removed from the above raw data the result is:

	Conditions			
Subjects	−0.40	0.60	−0.60	0.40
	−0.40	−1.40	2.40	−0.60
	−0.40	−2.40	0.40	2.40
	−2.15	2.85	−1.35	0.65
	3.35	0.35	−0.85	−2.85

These deviations represent the unreliability of the measurements (section 5.4.2). They are the deviations from the values expected on the basis of the **conditions** and **subjects** effects alone.

$$SS_{\text{reliability}} = (t_1^2 + t_2^2 + t_3^2 + \cdots)$$

is the appropriate formula. (This formula is set out as the within-conditions variance estimate in section 5.4.3.)

Use it to calculate $SS_{\text{reliability}}$.

(f) Complete the ANOVA summary table for this experiment using the above *SS* calculations and taking the rules for *df*s from section 5.5.

Complete the test of significance by deciding whether to reject H_0.

(g) Carry out the analysis of the above data using an appropriate computer package. Obtain:

(i) The mean scores in the four **conditions**.
(ii) A graphical display of the means in (i) as a bar chart or equivalent.
(iii) An analysis of variance summary table with *p*-value or equivalent indication of statistical significance.

Check that you are able to interpret the result of this analysis of variance.

(h) Identify from the analysis in (f) or (g) the estimate of the variance of the population of which the deviations tabulated in (e) are a sample.

(i) Does SS_{subjects} have any role in this analysis?

5.2 An experiment was carried out to compare the effects of three drugs on performance. The measure of performance was mean reaction time on a series of standardized tasks.

The five subjects who took part were a random sample from the relevant population. Each subject was measured under the influence of each of the drugs and a drug-free control condition.

The mean reaction times recorded by the experimenter were (Winer *et al.*, 1991):

Subject	Drug 1	Drug 2	Drug 3	Control
1	30	28	16	34
2	14	18	10	22
3	24	20	18	30
4	38	34	20	44
5	26	28	14	30

Subject means: 27.0, 16.0, 23.0, 34.0, 24.5
Condition means: 26.4, 25.6, 15.6, 32.0
Overall mean: 24.9

(a) Plot the conditions effects unique to the first two **subjects** on the same graph. Do they show a consistent pattern in the effects of the **conditions**? (Hint: use raw scores for the plot.)
(b) Obtain the **conditions** effects unique to each of the first two **subjects** as deviations from the **subjects'** means.
Using intuition, which of these two **subjects** would you say is most affected by the **conditions**?
(c) Express the overall **condition** and **subjects** effects as deviations from the appropriate mean and hence calculate their *SS*s. (The between-conditions and between-subjects *SS*s are set out in sections 5.4.3 and 5.4.5.)
(d) If $SS_{\text{reliability}} = 112.8$, complete the ANOVA summary table and the test of hypothesis for the **conditions** effect (sections 5.4.3 and 5.4.5.)
(e) What is the size of the **conditions** effect *SS* expressed as a proportion of the appropriate total?
(f) Carry out the analysis of the above data using an appropriate computer package. Obtain:
   (i) The mean scores in the four **conditions**.
   (ii) A graphical display of the means in (i) as a bar chart or equivalent.
   (iii) An analysis of variance summary table with *p*-value or equivalent indication of statistical significance.
   Check that you are able to interpret the result of this analysis of variance.

6 Two-factor independent groups design

6.1 INTRODUCTION

A more complete and detailed account of the design introduced in section 2.3 now follows in section 6.2. A new example is introduced which will serve to illustrate all aspects of the two-factor design. Questions posed in the language of the users of the findings of the experiment are set out at the start of section 6.2.3. The results are interpreted in the light of the mean scores obtained under experimental conditions and linked to the formal language of hypothesis testing.

6.2 EXAMPLE OF TWO-FACTOR DESIGN

Consider, as an example, an experiment from occupational psychology which aims to explore simultaneously the effects of type of sewing machine and method of training on the time needed by a machinist to sew a standard garment.

There are three alternative models of the industrial sewing machine. They form the levels of the **machine** factor. They will be referred to as *machine 1*, *machine 2* and *machine 3*.

There are four alternative training programmes for the machinists who operate the machines. They form the levels of the **training** factor. They will be referred to as *method 1*, *method 2*, *method 3* and *method 4*.

The number of **minutes** needed by a machinist to sew a standard garment is the dependent variable.

6.2.1 The conduct and layout of the example experiment

The experiment was conducted according to the requirements for generalization and proper experimental procedure, as discussed in sections 1.2, 1.3 and 3.3; namely, subjects were randomly selected from the appropriate population and were randomly allocated to a set of conditions.

Sixty employees were allocated, 5 to each of the 12 combinations of conditions.

The layout diagram is as in Table 6.1. It shows the numbers of subjects in each combination of conditions.

Table 6.1 Layout diagram for two-factor design example

		Training				
		Method 1	**Method 2**	**Method 3**	**Method 4**	Total
Machine	**1**	$n=5$	$n=5$	$n=5$	$n=5$	20
	2	$n=5$	$n=5$	$n=5$	$n=5$	20
	3	$n=5$	$n=5$	$n=5$	$n=5$	20
	Total	15	15	15	15	60

6.2.2 The mean scores from the example experiment

The results, expressed as the mean times for each group of five subjects for sewing up garments, are set out in Table 6.2. This table displays the mean times in the 12 cells. These are called **cell means**. Also displayed are a further seven means along the margins of the table. These are called **marginal means**. The marginal means are the means of all measurements in a particular row or column. In the case that there are identical numbers of subjects in each cell the marginal means are the means of the means in the particular row or column. This applies here.

Table 6.2 Mean sewing times (in minutes)

		Training				All
		1	**2**	**3**	**4**	
Machine	**1**	12.0	13.0	15.0	20.0	15.0
	2	9.0	8.0	12.0	15.0	11.0
	3	25.0	17.0	19.0	15.0	19.0
	All	15.33	12.67	15.33	16.67	15.00

In the bottom right-hand corner is displayed the **overall mean**. This is the mean of the measurements of all 60 subjects taking part in the experiment. It is the mean of all the row means and of all the column means provided there are identical numbers of subjects in each cell, as in this example.

6.2.3 Examples of users' questions

Several questions are of interest to the managers of the clothing factory. The managers are the users of the findings of the experiment. Their questions have led to money being spent on running the experiment. Examples of their questions are:

1. Does the method of **training** make a difference to the time taken sewing a standard garment?
2. Does the type of **machine** used make a difference?
3. If *machine 2* is to be used does it matter which method of **training** is used?
4. Does the method of **training** to be used depend on the choice of **machine**?

These questions fall into the standard categories of questions as discussed in section 2.3.1. They are, respectively,

1. The main effect of the factor **training**.
2. The main effect of the factor **machine**.
3. The simple effect of the factor **training** at the second level of the factor **machine**.
4. The interaction of the two factors.

For each question a set of means provides the basis for an answer. The sets of means are read off from a single row, column or margin of Table 6.2, except for the interaction (question 4), for which the answer is based on all three rows of means. There follows a list of the appropriate means:

1. 15.33 12.67 15.33 16.67
2. 15.0 11.0 19.0
3. 9.0 8.0 12.0 15.0
4.

		Training			
		1	2	3	4
Machine	1	12.0	13.0	15.0	20.0
	2	9.0	8.0	12.0	15.0
	3	25.0	17.0	19.0	15.0

The above sets of means only provide partial answers to the four questions. Consider question 1: 'Does the method of training make a difference to the **time taken** sewing a standard garment?' Examination of the means, with or without the aid of the bar chart in Fig. 6.1, suggests that garments are sewn most quickly by machinists who had experienced training *method 2*. They needed only 12.67 minutes on average compared to 15.33 minutes for the next quickest. This would imply that yes, the method of **training** does make a difference.

However, it is possible that the **training methods** are identical in their effects. If so, then the relatively small value of the mean sewing time in training *method 2* must be the result of sampling fluctuation. In other

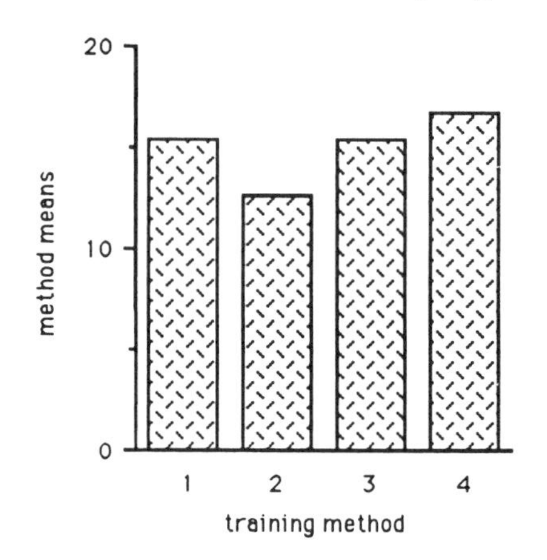

Fig. 6.1 Main effect of training.

words, the chance effects of random allocation resulted in the best machinists being in the group that experienced training *method 2*. Only a test of significance can lead to a decision.

A similar discussion can take place about questions 2 and 3. The appropriate set of means can be easily interpreted in the context of the question posed by the user of the findings of the experiment. This is not the case, however, for question 4, the interaction question. The next section discusses this.

6.3 THE EFFECT OF THE INTERACTION OF THE FACTORS

In order to clarify the concept of interaction in its relationship to the simple and main effects, there follows a re-examination of the example experiment taking the simple effects as the starting point.

As described in section 2.3, and referring again to the above example, the interaction of one factor with the other is by definition the existence of differences among the simple effects of one factor at the various levels of the other factor. How is this to be seen in the sewing machine example?

Consider the simple effect of the **training** factor at the first level of the **machine** factor. It is defined by the mean sewing times for the type 1 machine and the various training methods. The means are: 12.0, 13.0, 15.0 and 20.0. Expressed as deviations from the mean sewing time with this type of machine (i.e. 15.0) they are −3, −2, 0 and 5. It appears that training *method 4* is by far the least effective with this type of **machine**.

Likewise, the complete set of the simple effects of the **training** factor at the various levels of the **machine** factor are given in Table 6.3. Each set of deviations stands relative to the mean score of subjects experiencing the stated level of the **machine** factor.

Table 6.3 Simple effects of training

	Training
Machine type 1	(−3, −2, 0, +5)
Machine type 2	(−2, −3, +1, +4)
Machine type 3	(+6, −2, 0, −4)

As set out in Table 6.3, the simple effects can be seen to differ one from the other. The *machine type 3* simple effect of **training** differs markedly from the other two simple effects. In other words, the pattern of deviations corresponding to the four training methods appears very different for the third type of machine compared with the other two types. *Method 4* is the best training for *machine type 3*, whereas it is the worst for *machine types 1* and *2*. This is what is meant by interaction.

Definition of interaction

Interaction occurs when one factor's effect on the dependent variable shows a different pattern at the various levels of the other factor.

This is illustrated in Fig. 6.2, where the simple effects are displayed as bar charts of the means. The simple effect of **training method** for *machine 3* appears to differ even more from those for the other two machines when the three bar charts are superimposed to form Fig. 6.3. This is the conventional plot of the cell means used to display interaction. It is referred to as an **interaction diagram**.

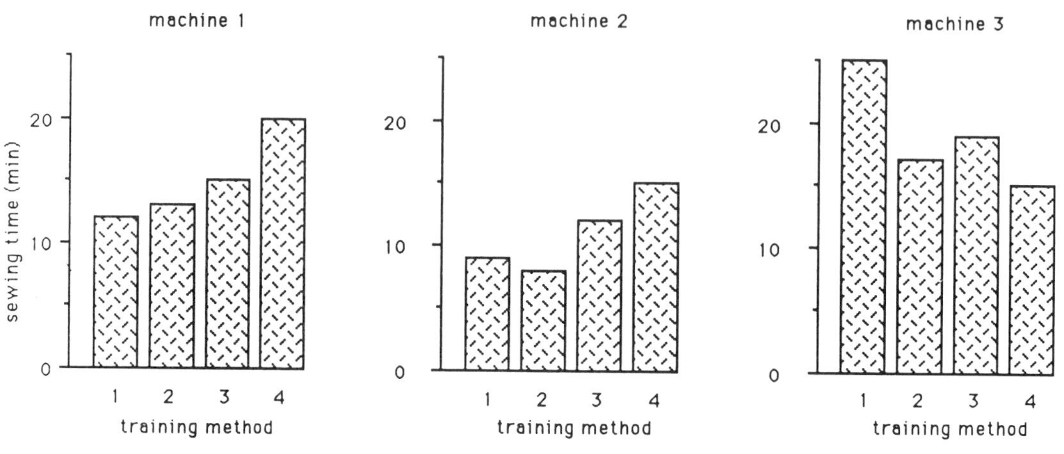

Fig. 6.2 Mean sewing times by machine and training.

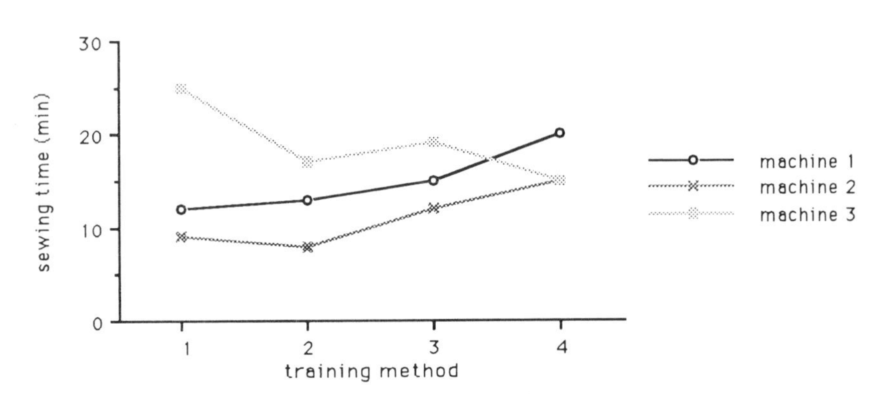

Fig. 6.3 Interaction diagram – machine × training.

The extent of departure of the three lines from the parallel indicates the extent of the interaction. If the lines are parallel there is no interaction because the same pattern is shown in each simple effect.

As previously discussed, any pattern among the means which suggests interaction could be the result of sampling fluctuation. A test of significance is required in order that a decision can be made.

How does the main effect relate to the simple effects? It is simply their average. This is easily illustrated for our numerical example. The means of the columns of Table 6.3 are 0.3333, −2.3333, 0.33333 and 1.6667. These are the deviations which make up the main effect of the **training** factor. They are obtained directly by subtracting 15.00, the overall mean, from, in turn, 15.33, 12.67, 15.33 and 16.67. Obtaining the same four deviations by two different routes lends credibility to the assertion that the main effect is the mean of the simple effects.

The pattern of the main effect of the **training** factor is dissimilar to any of its three component simple effects. This is a feature of interaction and its presence challenges the existence of the factors as meaningful concepts. The **training** factor is seen here as having a meaning critically dependent on subjects' exposure to another factor: **type of machine**.

This discussion could equally have taken place in terms of the simple effects of the **machine** factor at the various levels of the **training** factor. The same result would have been achieved. If one set of simple effects differs, so must the other set. In other words if factor 1 is interacting with factor 2 then factor 2 is interacting with factor 1.

6.4 THE PRINCIPLES OF THE ANALYSIS OF VARIANCE FOR THE TWO-FACTOR DESIGN

6.4.1 Introduction

In all essentials the analysis of variance for this design follows the principles for the single-factor independent groups design set out in section 4.2. The variance of scores on the untreated dependent variable (that is the between-subjects variance) is estimated within-groups and between-groups. The within-group estimate cannot be inflated by any effects of the treatment factors or interaction, whereas the between-groups estimate can.

The comparison of the within and between estimates in an F-test leads to a decision on the presence of between-group treatments effects.

Comparisons between means can provide estimates of the underlying population variance whether they arise from a main effect, a simple effect or an interaction.

The details of how these estimates are obtained follow. In general, the relevant means are reduced to deviations, squared, added and multiplied by a **weight** to give the SS. This was introduced in section 3.2. (The weight is the number of scores on which is based each of the relevant means.)

$$SS = \text{(weight) (sum of squared deviations among means)}$$

When the SS is divided by the appropriate degrees of freedom the mean square results. The mean square is the estimate of the between-subjects variance.

6.4.2 Variance estimate from a main effect

The variance of the means that constitute a main effect estimates the between-subject variance divided by the number of scores obtained under each condition. However, by identifying the number of scores obtained under each condition as the weight as introduced in section 3.2 and developed in section 6.4.1, the main effect means can be used to estimate the between-subjects variance.

For example, the SS of the means of **method of training** is obtained as the sum of squares of deviations of the means multiplied by the weight. The weight is 15. This is the number of subjects experiencing each **method of training** (Table 6.1).

SS of **method of training**
$= 15\ [(0.333)^2 + (2.333)^2 + (0.333)^2 + (1.667)^2]$
$= 126.67$

The corresponding degrees of freedom are $(4-1)=3$. Hence the MS, the estimate of the between-subjects variance from the variation among the means of the **training** factor is

$$MS = \frac{126.67}{3} = 42.22$$

The general symbolic form of the calculation follows. Suppose factor 1 has p levels and factor 2 has q levels. Suppose further that there are n subjects per group.

The SS for factor 1 (**type of machine**) is based on the p marginal means 15.0, 11.0 and 19.0, for which the weight is nq (i.e. (5) (4)). This is because each of the means 15.0, 11.0 and 19.0 is based on (5) (4) = 20 measurements.

Likewise the means that are relevant to the main effect of factor 2 (**method of training**) are the q marginal means each obtained from all np subjects whose scores contributed. The weight in this case is np. For factor 2 the relevant means are 15.33, 12.67, 15.33 and 16.67 and the weight is np, or (5) (3).

The degrees of freedom are always one less than the number of levels of the factor, i.e. $(p-1)$ for factor 1 and $(q-1)$ for factor 2.

A similar calculation for factor 1, the **type of machine**, involving degrees of freedom of $(p-1)=2$ and weight $nq=20$ leads to a between-subjects variance estimate value of 320.00.

6.4.3 Variance estimate from a simple effect

Consider one of the simple effects of the **method of training** factor displayed in the bar chart in Fig. 6.2. For example: the first one, the effect of the factor at the first level of the **machine** factor.

It is specified by the four relevant means: 12, 13, 15 and 20. The SS is obtained as the sum of squares of the deviations of these means from their common mean multiplied by the weight. The common mean is $(12+13+15+20)/4=15.0$. The weight is 5 since this is the number of scores on which each mean is based (Table 6.1).
The deviations are:

$12-15=-3$
$13-15=-2$
$15-15=\ \ 0$
$20-15=+5$

$SS\ = (5)[(-3)^2 + (-2)^2 + (0)^2 + (5)^2] = 190$
$MS = 190/3$
$\quad = 63.33$

(Note that $df=(4-1)=3$.) Hence the estimate of between-subjects variance obtained from one of the simple effects is 63.33.

6.4.4 Variance estimate from the interaction

Introduction

The discussion in section 4.2 is sufficient to explain the formation of variance estimates from main and simple effects. An extended further discussion is needed to explain the formation of a variance estimate from the interaction.

Two approaches to conceptualizing the interaction as deviations follow. Each uses the deviations to obtain the *SS* in the usual way.

SS and MS interaction derived from simple effects

Interaction of factor 1 with factor 2 has been defined in section 6.3 above as the variation among the simple effects of factor 2 at the various levels of factor 1, or vice versa.

The test of hypothesis of interaction is concerned with the question of whether an apparent interaction effect is solely due to the effect on the cell means of sampling fluctuation (i.e. H_0 is true) or due to a combination of the effects of sampling fluctuation and a real interaction effect (i.e. H_0 is false).

Consider, in the example, the variation of the simple effects of **training** at each of the three levels of **machine** from their mean. What is meant by the mean of several simple effects?

The concept only makes sense when the simple effects of **method of training** are expressed as deviations and when their mean is expressed as the set of deviations corresponding to the main effect. This was introduced in section 6.3 and is shown in Table 6.4. It is easily verified that the means of the columns set out along the lower margin of Table 6.4 are the deviations which describe the main effect of the **training** factor.

Table 6.4 Simple effects of training as deviations

	Training			
Machine 1	(−3,	−2,	0,	+5)
Machine 2	(−2,	−3,	+1,	+4)
Machine 3	(+6,	−2,	0,	−4)
Means of columns	(0.333,	−2.333,	0.333,	1.667)

The deviations which measure the amount by which the three simple effects differ from their mean are obtained (as usual) by subtracting the common mean from each of the items. This requires 0.333 to be subtracted from all items in the first column, −2.333 to be subtracted from all items in the second column, etc. The result is shown in Table 6.5. The array of deviations in

Table 6.5 describes the variation among the simple effects. (Note: if the three simple effects were identical, all 12 of these deviations would be zero.)

$$SS = (\text{weight})\ (\text{sum of squared deviations among means})$$

The weight is the number of scores on which is based any one of the relevant means. Here it is 5, the number of subjects in one cell (Table 6.1).

$$\begin{aligned} SS &= (5)[(-3.333)^2 + (0.333)^2 + \cdots + (-0.333)^2 + (-5.667)^2] \\ &= (5)\ (98.667) \\ &= 493.33 \end{aligned}$$

Table 6.5 Deviations of simple effects of **training**

Machine 1	(−3.333,	0.333,	−0.333,	3.333)
Machine 2	(−2.333,	−0.667,	0.667,	2.333)
Machine 3	(5.667,	0.333,	−0.333,	−5.667)

The mean square, which is the estimate of between-subject variance, is obtained, as usual, by dividing *SS* by *df*. The degrees of freedom are found by the same reasoning as that used in section 5.4.2, namely that since the deviations in every row and every column of the array in Table 6.5 must add to zero, only six of the twelve are free to be determined independently. The remainder are determined by the constraints of the row and column totals. The six free deviations determine the degrees of freedom value: $df = 6$.

The formula for this is $(p-1)(q-1)$. Hence

$$MS = \frac{493.33}{6} = 82.22$$

The estimate of the between-subjects variance from the interaction is therefore 82.22.

The same value could have been obtained from the variation among the simple effects of the other factor since interaction is symmetrically related to both factors.

SS and MS interaction derived from cell means

This same value for *SS* of interaction can also be obtained from the cell means directly using a different conceptualization. It is a similar approach to the one followed in section 5.4.2, where the effect of the conditions factor was removed from the balancing times data by subtracting the deviations that represent the main effect.

The rationale for this approach is based on an alternative definition of interaction.

Definition of interaction

Interaction is the variation among the cell means not due to the main effects of either conditions factor (in this case **method of training** and **type of machine**).

The definition is put into effect by removing the effects of first one factor and then the other to leave an array of means whose deviations can only represent the effect of interaction.

First remove the effect of the **type of machine** factor by subtracting its vector of deviations from the rows of the cell means array in Table 6.6. The means of the levels of the **machine** factor are 15, 11 and 19 with common mean 15. These reduce to deviations (0, −4, +4). Subtract these from all the means in respective rows (i.e. 0 from all means in the first row, −4 from all means in the second row, etc.) to obtain the array of means in Table 6.7. Notice that there is no variation among the row means in Table 6.7. The effect of the **machine** factor, which operates on the rows, has been removed.

Table 6.6 Mean sewing times (in minutes)

		Training				All
		1	**2**	**3**	**4**	
Machine	**1**	12	13	15	20	15.00
	2	9	8	12	15	11.00
	3	25	17	19	15	19.00
	All	15.33	12.67	15.33	16.67	15.00

Table 6.7 Mean times with effect of **machine** removed

		Training				Row means
Machine	**1**	12	13	15	20	15.00
	2	13	12	16	19	15.00
	3	21	13	15	11	15.00
Column means		15.33	12.67	15.33	16.67	

Next, the variation due to the **training methods** factor is removed by subtracting the effect of that factor in the form of a vector of deviations from the columns of the array of means in Table 6.7.

The **training methods** vector is (0.33, −2.33, 0.33, 1.67). Accordingly, 0.33 is subtracted from the values 12, 13 and 21 in the first column, −2.33 is subtracted from each of the means in the second column and so on to obtain the array of means displayed in Table 6.8.

The deviations of the cell means in Table 6.8 from their overall mean of 15 cannot be due to the effects of either **machine** or **training** factor. That

Table 6.8 Mean times with effects of **machine** and **training** removed

		Training				Row means
Machine	**1**	11.67	15.33	14.67	18.33	15.00
	2	12.67	14.33	15.67	17.33	15.00
	3	20.67	15.33	14.67	9.33	15.00
Column means		15.00	15.00	15.00	15.00	

these have both been fully removed is evident from the lack of variation among the row and column marginal means.

The remaining variation is therefore due to the effect of interaction of the two factors. It is taken for granted that sampling fluctuation is either an alternative or additional cause of the remaining variation.

The deviations from the overall mean of 15 of the adjusted cell means in Table 6.8 are given in Table 6.9, which is the same as Table 6.5. The two approaches lead to the same set of deviations to represent the effect of the interaction.

The $SS = 493.33$, as before, to give the estimate of between-subjects variance as $MS = 82.22$.

Table 6.9 Mean times with effects of **machine** and **training** and the overall mean removed

	Training			
Machine 1	−3.33	+0.33	−0.33	+3.33
Machine 2	−2.33	−0.67	+0.67	+2.33
Machine 3	+5.67	+0.33	−0.33	−5.67

Note that there is a convenient computational method for calculating *SS* for interaction, described in section 13.5.5.

6.5 THE SUMMARY TABLE AND TESTS OF SIGNIFICANCE

6.5.1 Rationale

There are three independent estimates of the between-subjects variance arising respectively from the effects of the factors and their interaction. These are unbiased estimates only if the apparent effects of the factors and interactions are caused by sampling fluctuation. Any true effects of the conditions factors and interaction will inflate the values of the estimates as discussed previously.

The *F*-test of significance depends on comparing the above estimates with an estimate that cannot be so inflated. Such an estimate is the within-groups estimate. It is exactly comparable to the estimate of the same name in the single-factor design, and is calculated by a similar formula (section 4.3.2). For the calculation see section 6.6.2.

As in the single-factor independent groups design, *SS* is a measure of variation so scaled that SS_{total}, the sum of squared deviations from the overall mean of every individual score can be decomposed into the four *SS*s which are the measures of variation due to the above sources.

The general expressions **factor 1** and **factor 2** will be used to stand for **type of machine** and **method of training** respectively.

It is an algebraic truism that:

$$SS_{\text{total}} = SS_{\text{factor 1}} + SS_{\text{factor 2}} + SS_{\text{interaction}} + SS_{\text{within}}$$

The variation between-subjects is also referred to as individual differences, residual, error or within-groups variation.

The degrees of freedom follow from the discussion above or directly from the formulae set out in the next section.

6.5.2 Formulae for degrees of freedom

df for $SS_{total}=N-1$, for $SS_{factor\,1}=p-1$, for $SS_{factor\,2}=q-1$, for $SS_{interaction}=(p-1)(q-1)$ and for $SS_{within}=pq(n-1)$, where N is npq, the total number of subjects, n is the number of subjects per group, and p and q are the number of levels (conditions) in factors 1 and 2 respectively.

6.5.3 The summary table and size of effect

The summary table

The summary table is shown in Table 6.10. The calculations of all *SS*s except SS_{within} and SS_{total} are demonstrated in sections 6.4.2, 6.4.3 and 6.4.4 in support of the explanation of principles. All are calculated in section 6.6.2 in the demonstration of the formulae for hand calculation.

Table 6.10 Analysis of variance summary table

Source	*df*	*SS*	*MS*	*F*
Factor 1 (Machine)	2	640.00	320.00	118.52
Factor 2 (Training)	3	126.67	42.22	15.64
Interaction	6	493.33	82.22	30.45
Within	48	129.38	2.70	
Total	59	1389.38		

Size of effects

The size of the effects of each factor and the interaction can be expressed as the proportion of the total *SS* explained by each.

Thus factor 1 (**machine**), factor 2 (**training**) and their interaction explain, respectively,

$$640.00/1389.38=46\%$$
$$126/1389.38 = 9\%$$
$$493.33/1389.38=36\%$$

of the total variation in sewing times. This leaves 9% of the variation unexplained.

6.5.4 Interpretation of the analysis

The main component of the interpretation is the testing of hypotheses about the main effects and interaction. For each of these sources of variance F is calculated in order to decide whether or not H_0 should be rejected. H_0 represents the hypothesis that the source of variation has no effect.

F is calculated by dividing the mean square of the source by the *MS* within. The logic of this is discussed in section 6.4. The between-subjects variance is estimated by MS_{within} as 2.70.

The critical values of F at the 0.05 significance level (to be found tabulated in Appendix F) for factor 1, factor 2 and interaction, respectively, are 3.23, 2.84 and 2.34 based on (2, 48), (3, 48) and (6, 48) degrees of freedom. Since these are greatly exceeded by the observed values of F, the H_0 of no effect can be rejected in every case.

The conclusion is that both factors and the interaction make a contribution to the variation in score.

Furthermore, since subjects were randomly allocated to groups it can be concluded that **training method**, **machine type** and their interaction all independently caused differences among the cell and marginal means. In other words, the factors and interaction affect the sewing times for the standard garment.

Interpretation of the user's questions (section 6.2.3)

The questions were:

1. Does the **method of training** make a difference to the **time taken** sewing up a standard garment?
2. Does the **type of machine** used make a difference?
3. If **machine 2** is to be used does it matter which **method of training** is used?
4. Does the **method of training** to be used depend on the choice of **machine**?

The answer to question 1 is 'yes', since 15.64 exceeds the critical F of 2.84.

The answer to question 2 is 'yes', since 118.52 exceeds the critical F of 3.23.

The answer to question 3 is found using the approach in section 6.4.3. It cannot be obtained directly from the ANOVA summary table. The answer is 'yes' since F for the simple effect is 18.52.

The answer to question 4 is 'yes', since 30.45 exceeds the critical F of 2.34.

6.6 COMPUTATIONAL FORMULAE FOR HAND CALCULATION OF SSs

6.6.1 Algebraic formulae

Most readers will prefer to obtain the summary table (Table 6.10) by use of one of the many statistical computer packages. However, the formulae for SSs that are most convenient for hand calculation are as follows:

$$SS_{\text{total}} = \Sigma\Sigma X^2 - \frac{T^2}{N}$$

$$SS_{\text{factor 1}} = \left(\frac{T_{1\cdot}^2}{nq} + \frac{T_{2\cdot}^2}{nq} + \cdots + \frac{T_{p\cdot}^2}{nq}\right) - \frac{T^2}{N}$$

$$SS_{\text{factor 2}} = \left(\frac{T_{\cdot 1}^2}{np} + \frac{T_{\cdot 2}^2}{np} + \cdots + \frac{T_{\cdot q}^2}{np}\right) - \frac{T^2}{N}$$

$$SS_{\text{within}} = \Sigma\Sigma X^2 - \left(\frac{T_{11}^2}{n} + \frac{T_{12}^2}{n} + \cdots + \frac{T_{pq}^2}{n}\right)$$

$$SS_{\text{interaction}} = SS_{\text{total}} - SS_{\text{factor 1}} - SS_{\text{factor 2}} - SS_{\text{within}}$$

where $\Sigma\Sigma X^2$ is the sum of squares of all raw scores, $T_{i\cdot}$ is the total of the scores at the ith level of factor 1, $T_{\cdot j}$ is the total of the scores at the jth level of factor 2, T_{ij} is the total of the scores at the ith level of factor 1 and the jth level of factor 2, T is the total of all scores, p and q are the number of levels of factors 1 and 2 respectively, n is the number of subjects per cell and N $(=npq)$ is the total number of subjects.

Degrees of freedom

df for $SS_{\text{total}} = N-1$, for $SS_{\text{factor 1}} = p-1$, for $SS_{\text{factor 2}} = q-1$, for $SS_{\text{interaction}} = (p-1)(q-1)$, and for $SS_{\text{within}} = pq(n-1)$.

An example will illustrate the calculation and the tests of significance.

6.6.2 Hand-worked numerical example of two-factor design

There follows a presentation of the 3 × 4 two-factor design referred to as the sewing times example of section 6.2. The complete array of measurements on the d.v. for the 60 subjects is shown in Table 6.11. The corresponding means are as set out in Table 6.2.

It is normally expected that a computer is used to calculate all *SS*s. However, for those special occasions on which a hand calculation is required, the formulae of section 6.6.1 will now be illustrated.

Table 6.11 Sewing times in minutes

		Levels of factor 2, **training**			
		1	*2*	*3*	*4*
Levels of factor 1 **type of machine**	1	12.5	13.1	15.5	20.3
		11.5	12.5	14.2	22.0
		13.9	11.0	13.9	18.5
		12.8	12.1	14.0	19.4
		9.3	16.3	17.4	19.8
	2	9.6	7.3	12.1	15.2
		11.4	8.1	13.6	15.0
		7.3	8.9	11.4	17.6
		8.9	9.2	10.0	14.1
		7.8	6.5	12.9	13.1
	3	25.8	17.4	19.4	14.3
		27.2	17.9	17.1	15.5
		24.3	15.8	16.5	17.0
		22.0	15.2	20.0	14.8
		25.7	18.7	22.0	13.4

First, calculate the cell totals:

$$
\begin{aligned}
T_{11} &= 12.5+11.5+13.9+12.8+9.3 = 60\\
T_{12} &= 13.1+12.5+11.0+12.1+16.3 = 65\\
T_{13} &= \quad \ldots \quad 75\\
T_{14} &= \quad \ldots \quad 100\\
T_{21} &= 9.6+11.4+7.3+8.9+7.8 = 45\\
T_{22} &= \quad \ldots \quad 40\\
T_{23} &= \quad \ldots \quad 60\\
T_{24} &= \quad \ldots \quad 75\\
T_{31} &= \quad \ldots \quad 125\\
T_{32} &= \quad \ldots \quad 85\\
T_{33} &= \quad \ldots \quad 95\\
T_{34} &= \quad \ldots \quad 75
\end{aligned}
$$

Next, calculate the row totals:

$$
\begin{aligned}
T_{1\cdot} &= T_{11}+T_{12}+T_{13}+T_{14} = 300\\
T_{2\cdot} &= T_{21}+T_{22}+T_{23}+T_{24} = 220\\
T_{3\cdot} &= \quad \ldots \quad = 380\\
T_{\cdot 1} &= T_{11}+T_{21}+T_{31} = 230\\
T_{\cdot 2} &= T_{12}+T_{22}+T_{32} = 190\\
T_{\cdot 3} &= \quad \ldots \quad 230\\
T_{\cdot 4} &= \quad \ldots \quad 250
\end{aligned}
$$

Finally, the overall total $T = T_{1\cdot} + T_{2\cdot} + T_{3\cdot} = 900$, and again, as a check, $T = T_{\cdot 1} + T_{\cdot 2} + T_{\cdot 3} + T_{\cdot 4} = 900$. p is the number of levels of factor $1 = 3$; q is the number of levels of factor $2 = 4$; N is total number of measurements $= 60$; and n is number of measurements per cell $= 5$. Then:

$$
\begin{aligned}
\Sigma\Sigma X^2 &= 12.5^2 + 13.1^2 + \cdots + 22.0^2 + 13.4^2 = 14\,889.4\\
SS_{\text{total}} &= 14\,889.4 - (900^2/60) = 1389.4\\
SS_{\text{factor 1}} &= (300^2/20 + 220^2/20 + 380^2/20) - (900^2/60) = 640.0\\
SS_{\text{factor 2}} &= (230^2/15 + 190^2/15 + 230^2/15 + 250^2/15) - (900^2/60)\\
&= 126.67\\
SS_{\text{within}} &= 14\,889.4 - (60^2/5 + 65^2/5 + \cdots + 95^2/5 + 75^2/5)\\
&= 129.38
\end{aligned}
$$

Since:

$$SS_{\text{factor 1}} + SS_{\text{factor 2}} + SS_{\text{interaction}} + SS_{\text{within}} = SS_{\text{total}}$$

we can rearrange to get

$$
\begin{aligned}
SS_{\text{interaction}} &= SS_{\text{total}} - SS_{\text{factor 1}} - SS_{\text{factor 2}} - SS_{\text{within}}\\
&= 1389.4 - 640.0 - 126.67 - 129.38\\
&= 493.35
\end{aligned}
$$

This completes the calculation of all SSs required for the analysis of variance summary table in section 6.5.3.

6.7 UNDERLYING MODEL AND ASSUMPTIONS FOR TESTS OF SIGNIFICANCE

Underlying model

The effects of the conditions and interactions are assumed to be additive. That means that a particular level of a factor (i.e. a condition) is supposed to increment or decrement the scores of all subjects exposed to it by an identical amount.

This additive property implies that a particular subject's score is explainable by the following formula:

$$\begin{array}{l}\text{Expected}\\ \text{score for}\\ \text{a randomly}\\ \text{sampled subject}\end{array} = \text{overall mean} + \begin{array}{c}\text{effect of}\\ \text{factor 1}\end{array} + \begin{array}{c}\text{effect of}\\ \text{factor 2}\end{array} + \begin{array}{c}\text{effect of}\\ \text{interaction}\end{array}$$

For the numerical example above the model appears as

$$\begin{array}{l}\text{Expected}\\ \text{score for}\\ \text{a randomly}\\ \text{sampled}\\ \text{subject}\end{array} = 15 + \left\{\begin{array}{r}0\\ -4\\ 4\end{array}\right\} + \left\{\begin{array}{r}0.333\\ -2.333\\ 0.333\\ 1.667\end{array}\right\} + \left\{\begin{array}{rrrr}-3.333 & 0.333 & -0.333 & 3.333\\ -2.333 & -0.667 & 0.667 & 2.333\\ 5.667 & 0.333 & -0.333 & -5.667\end{array}\right\}$$

The four terms in the above model when appropriately added give the value of the score expected when a randomly sampled subject experiences a particular combination of levels of the two factors. This is a direct extension of the model for the single-factor design set out in section 4.4.2.

According to this model the score for a subject experiencing the second level of factor 1 and the third level of factor 2 is

$$15+(-4)+(0.333)+(0.667)=12.0$$

Note that 12.0 is exactly the mean score of subjects experiencing the specified levels of factors 1 and 2. This shows that this model completely explains the observed cell means in terms of three independent effects – the two factors and their interaction.

Assumptions for validity of the F-test

The assumptions on which depends the validity of the tests of significance are: the use of random sampling and allocation; the additive nature of the conditions, and the scores having a normal distribution in the population. See section 4.6 for a fuller discussion of these assumptions.

6.8 EXERCISES

6.1 Consider the numerical example in section 6.2. The experiment is from occupational psychology. Factor 1 refers to three alternative designs for an industrial sewing machine. Call this factor **machine**. Factor 2 refers to four alternative training programmes for operatives of the machines. Call this factor **training**.

Suppose the dependent variable is the number of minutes needed by an operative to sew up a standard job.

Sixty employees were allocated at random, 5 to each of the 12 combinations of conditions. The mean times were as in Table 6.2.

(a) Identify the best and worst combinations of **machine** and **training** programme.
(b) Overall which is the worst **machine**? Which is the worst **training** programme?
(c) Display all the simple effects of the **training** factor as bar charts. Are there indications of an interaction? (See Fig. 6.2.)
(d) Which of the four simple effects of **machine** is the least like the main effect of **machine**?
(e) Express the main effect of **machine** as deviations. Do the same for the main effect of **training**.
(f) Subtract from every row of the array of cell means the corresponding deviation due to the main effect of **machine**. This removes the effect of **machine** from the data (and leaves every row having the same mean) (Table 6.7).

Further simplify the data by removing the effect of **training** from the columns (thus leaving every column having the same mean). (The result is same as Table 6.8.)

What is the explanation for this remaining variation among the means in this twice-simplified table?
(g) Analyse this experiment by use of a suitable computer system. Obtain:

(i) The mean scores that express both main effects and the interaction.
(ii) A graphical display for each main effect and the interaction.
(iii) An analysis of variance summary table with p-value or equivalent indication of significance of each of the three effects referred to in (i) and (ii) above.

Compare the results with Table 6.10 and the interpretation of the analysis in section 6.5.4.

6.2 Thirty male and thirty female gerbils were randomly allocated, in single sex groups, to one of the following three conditions:

no experience of novel food
smell of novel food
contact with demonstrator gerbil used to novel food

There were 10 gerbils in each group. The dependent variable was the amount of the novel food eaten in grammes.

The mean scores in each group were:

		Condition			
		1	2	3	Means
Sex	Female	0.25	0.21	0.32	0.26
	Male	0.23	0.15	0.16	0.18
	Means	0.24	0.18	0.24	0.22

The ANOVA summary table was:

Source	*SS*	*df*
Condition	0.048	2
Sex	0.096	1
Interaction	0.052	2
Within (error)	0.342	54

(a) Complete the tests of significance of **condition**, **sex** and **condition** × **sex** interaction. (Table 6.10 is a model.)
(b) Display all three sources of variation on appropriate sketch graphs.
(c) What is the difficulty of interpretation of the main effect of **condition**?
(d) Which is the largest simple effect of **sex**?
Which is the largest simple effect of **condition**?
(e) Complete the tests of significance of each of the simple effects of **condition** (section 6.4.2).
(f) Are any of the simple effects of **sex** not significant?
(g) How do the sizes (in terms of *SS*s) of the simple effects of **condition** compare with the sizes of the main effect of **condition** and the interaction? Do you notice anything strange?
(h) Summarize the results of the experiment in simple English.
(i) Given that the deviations which describe each of the main effects and interaction are as set out in the additive model:

$$\text{amount eaten} = 0.22 + \underbrace{\begin{Bmatrix} +0.04 \\ -0.04 \end{Bmatrix}}_{\textbf{sex}} + \underbrace{\begin{Bmatrix} +0.02 \\ -0.04 \\ +0.02 \end{Bmatrix}}_{\textbf{condition}} + \underbrace{\begin{Bmatrix} -0.03 & -0.01 & +0.04 \\ +0.03 & +0.01 & -0.04 \end{Bmatrix}}_{\text{interaction}}$$

(i) What amount would you expect to be eaten by a male gerbil in the 'smell' condition? (See section 6.7.)
(ii) Use the deviations that represent the interaction to confirm the value $SS = 0.052$.

7 Single-factor independent groups design with covariate

7.1 INTRODUCTION

The use of a covariate to adjust the scores arising from an experiment was introduced in section 2.4. Its purpose is to improve the sensitivity and efficiency of the design. It will achieve this provided the covariate possesses certain properties.

The concept and technique of **adjustment** is dealt with in section 7.2 through an example using real data. Both an intuitive rule of thumb approximation and an exact method based on the technique known as **regression** are used.

Section 7.3 deals with the application of adjustment to the dependent variable of a single-factor independent groups experiment. The effect of adjustment on the analysis of variance summary table is demonstrated using the example that was introduced in section 2.4.

Section 7.4 shows how the covariate adjustment has an effect which appears as an additional term in the underlying model.

7.2 THE CONCEPT AND TECHNIQUE OF COVARIATE ADJUSTMENT

Covariate adjustment makes use of one measurement, the covariate, to adjust another measurement, the dependent variable. Both measurements must be continuous variables on scales with the equal value interval property (section 1.1.2). They need to be correlated for any benefit to be obtained. The magnitude of the correlation required for a given degree of benefit is discussed in section 7.4. The covariate is not the focus of interest in the research.

The idea is a simple one. Each subject's score on the dependent variable is adjusted to what it would have been if all subjects had an identical value of the covariate. An example will help to make this clear.

Suppose reading scores have been obtained for all pupils in a primary school class. The pupils' ages range from 7 years 2 months to 8 years 1 month. Figure 7.1 shows the relationship of **reading score** to **age**. A straight line has been drawn on the graph to represent the way the **reading scores** change with **age** over the whole group.

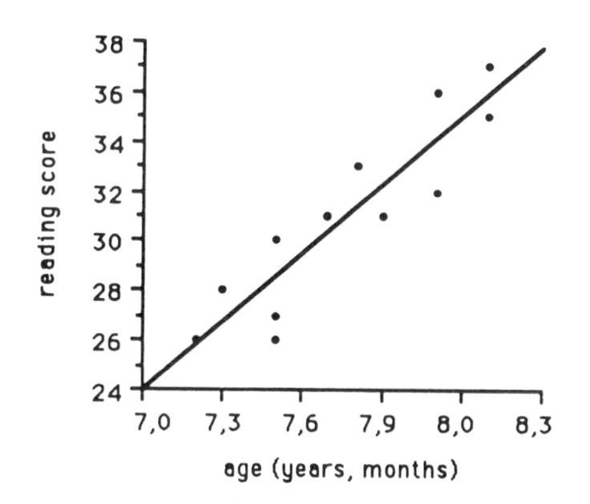

Fig. 7.1 Reading score versus age for a primary school class.

In this example, covariate adjustment would consist in estimating **reading scores** for the hypothetical situation in which all pupils had been measured at the same **age**. The estimation is based on the straight line relationship of **reading score** to **age** evidenced in Fig. 7.1. Scores increase from about 27 at 7 years 4 months to about 37 at 8 years 1 month. That is an increase of 10 units in 9 months or approximately 1 unit per month. This rate of increase corresponds to the **gradient** of the line on the graph in Fig. 7.1.

Suppose each individual progresses in reading at the rate shown by the whole group. In other words, suppose each individual advances by 1 unit per month. This supposition could only be confirmed by following the progress of a group of pupils over a period of time.

It is a straightforward matter to estimate what each pupil's **reading score** would have been at any given **age** by adding or subtracting 1 unit of score for each month of age added or subtracted.

Choose an arbitrary standard age, say 7 years 6 months, and adjust each pupil's score to this standard. The calculations are set out in Table 7.1. The adjusted scores range from 27 to 31 compared to 26 to 37 in the unadjusted scores. This shows how the variation in **reading scores** among the children has been reduced or **accounted for** or **explained by** the covariate **age**. This is seen also in the scatter-plot of adjusted **reading score** against **age** in Fig. 7.2 in comparison with Fig. 7.1.

Table 7.1 **Reading scores** and **ages** of 12 pupils

Pupil no.	*1*	*2*	*3*	*4*	*5*	*6*	*7*	*8*	*9*	*10*	*11*	*12*
Age	7,2	7,3	7,5	7,5	7,5	7,7	7,8	7,9	7,11	7,11	8,1	8,1
Reading score	26	28	26	27	30	31	33	31	32	36	35	37
Adjusted reading score	30	31	27	28	31	30	31	28	27	31	28	30

The rough and ready adjustment method used above is normally replaced by a mathematically rigorous method based on the best fitting straight line obtained by the regression technique. In the example above, using a suitable computer package, the line is found to be described by the following equation:

$$\text{score} = (-52.7) + (0.91)\ (\text{age in months})$$

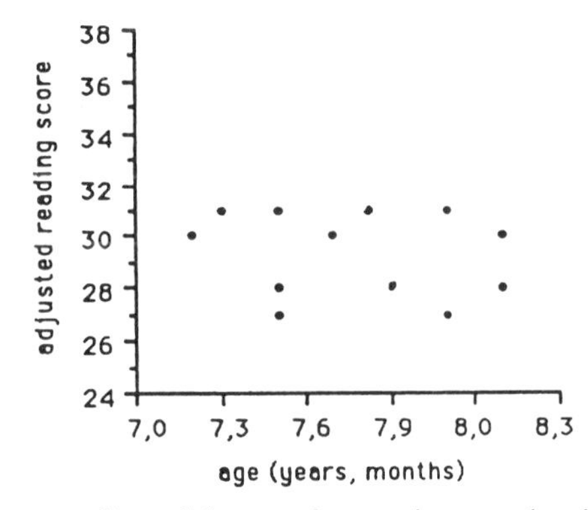

Fig. 7.2 Reading score adjusted for age for a primary school class.

This is a model which states that score increases by 0.91 units for each month's increase in age (i.e. the gradient is 0.91). This compares to 1.00 units gauged by eye. The formula, based on this value of the gradient, which generates the adjusted score value for a particular pupil is:

adjusted score = unadjusted score − (0.91) (actual age − standard age)

The use of this formula is now illustrated for pupil number 1:

$$\begin{aligned} \text{adjusted score} &= 26-(0.91)\,(86-90)\\ &= 26-(0.91)(-4)\\ &= 26+3.64\\ &= 29.64 \end{aligned}$$

(Note that 86 and 90 months are 7 years 2 months and 7 years 6 months, respectively.) This compares to 30 using the intuitive method.

7.3 THE EFFECT OF COVARIATE ADJUSTMENT ON VARIANCE ESTIMATES

The basic single-factor design is discussed in sections 2.1 and 4.1. It involves randomly allocating subjects to different levels of a factor and measuring a response, known as the dependent variable, for each subject.

In section 4.4 it was shown that the variation from subject to subject in the dependent variable (total *SS*) can be explained by a combination of individual differences between-subjects (within-group *SS*) and the differential effects of the conditions (between-groups *SS*). In fact:

total *SS* = between-groups *SS* + within-groups *SS*

Since the dependent variable values are altered by the covariate adjustments, all three of the *SS*s are affected. The result is an adjusted summary table and a correspondingly altered significance test result.

The example from section 2.4 will illustrate this. Eighteen rats are randomly allocated, nine to each of two conditions. Each group experiences a different drug treatment, **A** for the first group, **B** for the second. Scores are obtained on a dependent variable, **pulse**, under the influence of the drug. Scores are also obtained or are already available on a covariate. The **weight** of the rat in grams is the covariate. The scores are set out in Table 7.2.

Table 7.2 **Pulse** rates and **weights** of 18 rats

Rat No.	Drug	Pulse	Weight	Rat No.	Drug	Pulse	Weight
1	A	330	460	10	B	330	450
2	A	290	450	11	B	310	440
3	A	285	380	12	B	300	408
4	A	280	370	13	B	270	445
5	A	275	420	14	B	260	380
6	A	270	375	15	B	245	425
7	A	270	350	16	B	240	380
8	A	260	365	17	B	235	320
9	A	245	355	18	B	220	330
	Mean	278.3			Mean	267.8	

Consider the analysis of variance which results if no use is made of the **weights** of the rats. Table 7.3 summarizes the analysis.

Table 7.3 Analysis of variance on pulse rates

Source	*df*	*SS*	*MS*	*F*	*p-value*
Drug	1	501	501	0.51	0.487
Within-groups	16	15 856	991		
Total	17	16 357			

The estimate of the within-groups (individual differences or between-subjects) variance is 991. It is labelled 'within-groups *MS*' in the table. The interpretation of the *p*-value, as introduced in section 3.6, is that the effect of the **drug** is not significant since p is greater than 0.05.

If the **pulse** rates are adjusted for the **weights** of the rats a separate analysis of variance results. This is called an **analysis of covariance** (ANCOVA) and is summarized in Table 7.4. Note that the degrees of freedom for within-groups and total are each reduced by one for the adjusted model.

Table 7.4 Analysis of covariance for pulse rates

Source	*Adj. df*	*Adj. SS*	*MS*	*F*	*p-value*
Drug	1	844.1	844.1	1.97	0.181
Within-groups	15	6442.9	429.5		
Total	16	7287			

The estimate of the within-groups variance after adjustment is 429.5. This is about half the value obtained when no adjustment is made. The interpretation of this reduction is that the unadjusted variance is larger because it includes variation in **pulse** rates due to the variation in **weights** of the rats. 429.5 is the value that would be expected if rats of identical weights had been sampled.

Another expression for this adjustment procedure is **partialling out**. The effect of body **weight** has been partialled out. The benefit of adjustment can be seen by comparing the two summary tables. Because of the reduction in the within-groups mean square the F-value for the test of significance has increased from 0.51 to 1.97. In this example, since F_c at the 0.05 level is 4.54, the increase has not been large enough to lead to a decision to reject H_0. However, in other circumstances such an increase could have the effect of changing a non-significant result into a significant result.

Another way of viewing the benefit is by comparing the proportion of the total *SS* explained by the **drugs** treatments in the two analyses. This is set out in Table 7.5.

Table 7.5 Proportions of total *SS* explained in unadjusted and adjusted analyses

	Drugs *SS*	Total *SS*	Proportion
Unadjusted	501	16 357	3%
Adjusted	844	7287	12%

The benefit in terms of improved sensitivity and efficiency for this example is discussed in section 9.2.

Most computer programs for analysis of covariance provide the adjusted values of the means of the dependent variable. In an experiment conducted according to the proper procedures, subjects will be randomly allocated to groups. This should result in approximately the same distribution of values of the covariate in all groups. In turn, this should lead to only a very limited effect of adjustment on the differences among the means of the dependent variable. In the experiment illustrated here the relevant means are set out in Table 7.6. This shows that randomization did indeed share out the body **weights** fairly between the two groups. This has led to very little effect of adjustment on the differences among the mean **pulse** rates. In fact the adjusted difference is $279.9 - 266.2 = 13.7$ compared to the unadjusted difference, $278.3 - 267.8 = 10.5$. The slightly larger mean difference in the adjusted analysis has led to a corresponding increase in the value of the *SS* for the conditions factor. It increases from 501 to 844 due to the adjustment.

Table 7.6 Means in unadjusted and adjusted models

	Covariate mean (body **weight**)	Dependent variable mean	
		Unadjusted	Adjusted
Drug A	391.7	278.3	279.9
Drug B	397.6	267.8	266.2

7.4 UNDERLYING MODEL AND ASSUMPTIONS FOR TESTS OF SIGNIFICANCE

The underlying model of ANCOVA is based on the one for the single-factor independent groups design (section 4.6), but with the addition of a term which represents the effect of the covariate. It is:

Expected score for a randomly sampled subject	=	overall mean	+	conditions effect	+	covariate effect

For the numerical example this model appears as:

$$\text{expected score} = 273.05 + \begin{Bmatrix} +6.85 \\ -6.85 \end{Bmatrix} + 0.5389\,(\text{covar} - 394.61)$$

The effect of the factor is expressed as deviations, as in the other models introduced in Chapters 4, 5 and 6.

The **covariate effect** term in the model needs some explanation. The gradient value, 0.5389, has been obtained using a statistical computer package. The bracketed expression consists of **covar**, the value of an individual's covariate score with 394.61 subtracted from it. 394.61 is the mean covariate (i.e. the mean rat weight) over all 18 rats.

The rationale for the covariate effect term is that it is a deviation from the score that would be expected from a knowledge of only the overall mean and conditions effect. It is the amount by which the individual differs in covariate value from the overall mean covariate value multiplied by the scale factor (0.5389). The scale factor or gradient converts units of covariate into units of score.

(Note that in the model the conditions effect has been adjusted for the effect of the covariate. The deviations ± 6.85 are derived from the adjusted rather than the unadjusted means. Note also that the model is satisfied by all points on the plotted regression lines in Fig. 2.9. In mathematical jargon it is said to be the equation of those lines.)

The interpretation of this model is easily illustrated with an example. For a rat given **drug A** (i.e. the first level of treatment) and with a value of 400 grams for the covariate, the **pulse** rate is expected to be

$$\begin{aligned}\text{score} &= 273.05 + 6.85 + 0.5389(400 - 394.61)\\ &= 282.805\end{aligned}$$

All the usual assumptions for the analysis of variance model apply, namely that the conditions effect is additive and that the individual scores are normally distributed with identical variance in the populations from which the individuals are sampled.

There are the additional assumptions that the gradients of the regression lines are identical in the populations and that the regression effects and conditions effects are additive.

Furthermore, adjustment has to be credible. It is assumed, as mentioned in sections 2.4 and 7.2, that the scores of individuals can be estimated or adjusted to what they would have been if those individuals had had some other value of the covariate.

7.5 EXERCISES

7.1 A study compared three methods of teaching map reading:

Method I: all taught in the classroom
Method II: half taught in classroom, half in the field
Method III: all taught in the field

It was thought that the more formal education a person had experienced, the better he or she could be expected to do on the course. Accordingly, the number of years of post-16 education the person had experienced was used in the analysis as a continuous covariate (X).

The dependent variable was the person's score out of 10 on a test of map reading ability (Y).

Individuals from a sample of 21 were selected at random and allocated, seven to each method.

The means of Y and X were:

	Method 1	**Method 2**	**Method 3**	Overall
mean Y	4.43	7.57	6.71	6.24
mean X	2.14	3.43	2.71	2.76

The unadjusted and adjusted SSs for **methods** and within-groups are:

Source	SS	SS_{adj}
methods	36.95	16.94
within-groups	26.86	10.30

(a) Set out the ANOVA summary table for the effect of **method** based on the SSs unadjusted for, and again adjusted for, the effect of the covariate. Complete the test of significance in each case (section 7.3).

(b) If the underlying model is

$$\text{estimated score} = 6.24 + \begin{Bmatrix} -1.33 \\ +0.79 \\ +0.53 \end{Bmatrix} + 0.743(\text{covar} - 2.76)$$

obtain the estimated score for a person taught by Method I who had four years of post-16 education.

(c) Calculate and compare the proportions of variation in score explained by **methods** with and without the covariate (X) partialled out.

(d) Carry out an ANCOVA analysis of the data in Table 7.2 using an appropriate computer package. Obtain:
   (i) The adjusted and unadjusted mean scores on the dependent variable in the three conditions.
   (ii) The gradient.
   (iii) A graphical display of the means in (a) as a bar chart or equivalent.
   (iv) Analysis of variance summary tables for the adjusted and unadjusted *SS*s with *p*-values or equivalent indication of significance.

8 Contrasts and comparisons among means

8.1 INTRODUCTION

The analysis of variance and F-tests introduced in section 3.7 and discussed in detail in sections 4.4, 5.4 and 6.5 serve to test the **omnibus** or **general hypothesis** of differences among the population means. This section is concerned with more specific tests among the means.

Section 8.2 deals with the use of coefficients to formulate linear contrasts which specify the comparisons required by the researcher. The method for carrying out the test is also covered.

Section 8.3 deals with *a posteriori* tests of comparisons and section 8.4 presents a decision chart to assist in choosing the appropriate method.

8.2 FORMULATING AND TESTING A COMPARISON AMONG MEANS

8.2.1 Introduction

Take as an example the between-subjects gerbil experiment of section 4.3.1. The mean scores of the three groups of gerbils are set out in Table 8.1 and displayed graphically in Fig. 8.1.

The standard analysis of variance presented in Chapter 4 is testing the hypothesis that, in the population, the means do not differ against the alternative hypothesis that they do differ. The experimenter may wish to test more specific alternative hypotheses.

Table 8.1 Mean percentage returns

Degree of interruption	Mean	Group
None	38.75	1
Partial	60.63	2
Complete	47.00	3
Overall mean:	48.79	

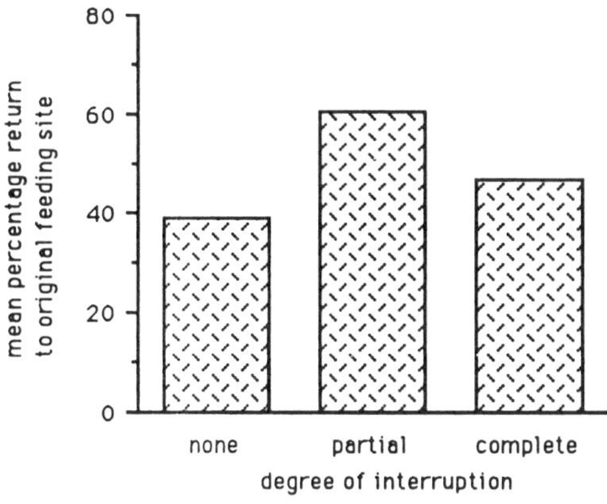

Fig. 8.1 Effect of interruption on the return to original feeding site.

Examples of specific alternative hypotheses for the gerbil experiment are:

(a) **None** differs from **Complete**.
(b) **Partial** differs from **Complete**.
(c) **None** differs from the mean of **Partial** and **Complete**.
(d) There is a trend to increasing mean scores in the population from **None** to **Partial** to **Complete**.

The corresponding null hypotheses state that the specified means do not differ or that there is no trend. All of (a) to (d) consist of comparisons (also known as contrasts) among conditions.

Contrasts which involve only two means, such as (a) and (b) above, are referred to as **pair-wise contrasts**, whereas contrasts involving more than two means, such as (c) and (d) above, are referred to as **multi-mean contrasts**.

In order to carry out the test of hypothesis the contrast is first specified in terms of **coefficients**. The coefficients are a set of integers selected according to rules set out in section 8.2.2.

The examples of specific hypotheses with the appropriate coefficients are:

(a) **None** differs from **Complete**.
   coefficients $(-1, 0, +1)$
(b) **Partial** differs from **Complete**.
   coefficients $(0, -1, +1)$
(c) **None** differs from the mean of **Partial** and **Complete**.
   coefficients $(-2, +1, +1)$
(d) There is a trend to increasing mean scores from **None** to **Partial** to **Complete**.
   coefficients $(-1, 0, +1)$

8.2.2 The rules for selecting coefficients

1. There must be a coefficient associated with each mean.
2. Two sets of means are identified. One set of means is to be contrasted with another set. The coefficients of the means in a set are the same (except for trend coefficients. See 5). The coefficients in one set always differ from those in the other set in sign and may differ in magnitude.

3. The coefficients add to zero.
4. Means not included in the contrast have zero coefficients.
5. A trend across a particular ordering of the means is specified by coefficients which increase in equal sized steps from left to right, e.g.

$$\begin{array}{ccccc} -1 & 0 & +1 & & \\ -3 & -1 & +1 & +3 & \\ -2 & -1 & 0 & +1 & +2 \end{array}$$

etc., depending on the number of means. Thus a trend is a comparison between the weighted means with positive and those with negative signed coefficients.

8.2.3 Calculation of the linear contrast function

The **linear contrast function** of the means is calculated by multiplying each mean by its coefficient and totalling the products. Consider hypothesis example (c) above:

$$L=(-2)(38.75)+(+1)(60.63)+(+1)(47.00)=30.13$$

Here L stands for the linear contrast function. The numerical value of L is used in a formula to obtain the sum of squares for the contrast. This is the sum of squares (SS) needed for completion of the test of significance.

8.2.4 The sum of squares of the contrast

The SS for the contrast is the square of the difference between two weighted means, the mean of the means with negative coefficients and the mean of the means with positive coefficients.

Fortunately the formula is very easy to use. It is:

$$SS=\frac{nL^2}{c_1^2+c_2^2+c_3^2+\cdots}$$

where n is the sample size for any one treatment or condition and c_1, c_2, c_3, etc. are the coefficients.

In this example we obtain

$$SS=\frac{(8)(30.13)^2}{2^2+1^2+1^2}=1210.42$$

8.2.5 The test of significance of the contrast

All contrasts have one degree of freedom. Hence the mean square of the contrast has the same value as the SS.

The F-test is constructed by dividing the MS of the contrast by the MS_{error} that was used in the general ANOVA F-test. In the gerbil example the ANOVA summary table for the general F-test of conditions is set out in Table 8.2 (reproduced from Table 4.3).

Table 8.2 Summary table for analysis of variance

Source of variation	*SS*	*df*	*Mean square*	*F*
Between-groups	1953	2	976.50	4.51
Within-groups (error)	4545	21	216.42	
Total	6498	23		

Continuing with the test of hypothesis of contrast example (c) above:

$$F = \frac{MS \text{ of contrast}}{MS_{\text{error}}} = \frac{1210.42}{216.42} = 5.593$$

This F value, 5.593, has to be compared with the appropriate critical value of F with (1, 21) degrees of freedom which is 4.32 at the 0.05 significance level. Note that the *df* of the MS_{error} is taken directly from the analysis of variance.

Since F exceeds its critical value a decision is made to reject H_0 and conclude that **None** differs from the mean of **Partial** and **Complete**.

The calculations for examples (a), (b) and (d) follow the same pattern. Only the coefficients are different.

8.2.6 The directional test of significance of a contrast

In most cases the researcher wishes to test a hypothesis of a contrast which has a particular direction. For example, consider again the example (b):

(b) **Partial** differs from **Complete**.

The researcher may have been interested in testing the directional hypothesis:

(b1) **Partial** exceeds **Complete**.

The tables of critical F usually available do not provide the appropriate values for directional tests. However, since there is only one degree of freedom for the numerator (always true for contrasts), the F value is the square of the corresponding t value. The tables of critical t values (Appendix F.1) provide directional values. The directional critical t for 21 degrees of freedom at the 0.05 significance level is 1.721. The square of this, 2.962, is the directional critical F.

To complete the directional test it is only necessary to compare the appropriately calculated F value with 2.962. When it exceeds its critical value, the decision is made to reject H_0 in favour of H_1.

(Note that rejection is ruled out immediately if the observed values of the means are not in the direction specified by the alternative hypothesis. Note also that for the 0.05 significance level the directional critical F can be obtained directly from Table F.2 in Appendix F.)

8.2.7 Testing a contrast for a within-subjects design

Take as an example the within-subjects dowel balancing experiment in section 5.4.1 (Table 8.3). Suppose it is required to test the contrast:

silent versus the mean of **speaking** and **humming**

Table 8.3 Mean balancing times

Condition	Mean	Group
Silent	10.0	1
Speaking	6.0	2
Humming	5.0	3
Overall mean:	7.0	

The coefficients are -2, $+1$ and $+1$.

$$L = (-2)(10.0) + (+1)(6.0) + (+1)(5.0) = -9.0$$

$$SS = \frac{(4)(-9.0)^2}{2^2 + 1^2 + 1^2} = 54.0$$

$$F = \frac{54.0}{4.0} = 13.5$$

on (1, 6) degrees of freedom.

Here 4.0 is the $MS_{\text{reliability}}$ from Table 5.4. It is the MS_{error} used in the general F-test and its degrees of freedom are 6.

The critical F is 5.99 at the 0.05 significance level. Since 13.5 exceeds 5.99 it is decided to reject H_0 and conclude that **silent** differs from the mean of **speaking** and **humming**.

8.3 *A POSTERIORI* TESTS OF COMPARISONS

8.3.1 Introduction

An *a priori* test of contrast is one that the experimenter had decided on when the experiment was designed. The researcher had a theory which predicted the contrast. The techniques described in section 8.2 apply to *a priori* contrasts.

An *a posteriori* (also known as *post hoc*) test of a contrast is one that the experimenter decides on after seeing the results. A theory is generated to explain a specific contrast that interests the researcher. A more stringent criterion for rejection of the null hypothesis is required because contrasts showing, possibly by chance, large differences between means are more likely to attract the attention of the researcher.

Two techniques will be described. The Scheffé corrected F-test for *a posteriori* multi-mean comparisons and the Newman–Keuls test for *a posteriori* pair-wise comparisons.

8.3.2 Scheffé corrected *F*-test

This is the most widely used of a number of well-known *a posteriori* tests. A full account can be found in Scheffé (1959).

The procedure is the same as for the uncorrected F-test described in section 8.2 above, but makes use of a corrected critical F. The corrected critical F is obtained by looking up the critical value of F (known as F_c) using the same degrees of freedom as used in the general F-test of conditions and multiplying it by $(k-1)$, where k is the number of different conditions included in the experiment.

The corrected F_c is larger than the uncorrected F_c. This has the effect of requiring a larger difference between means for rejection of H_0 in the *a posteriori* comparison.

Example of the Scheffé corrected F

Consider the *a posteriori* test of example (c) for which the observed value of F is calculated to be 5.593:

(c) **None** differs from the mean of **Partial** and **Complete**.

It is necessary to use (2, 21) rather than (1, 21) degrees of freedom in looking up the critical F. This gives 3.47, at the 0.05 level of significance, which value is then multiplied by $(3-1)$ to give $(2)(3.47)=6.94$.

The observed F value of 5.593 does not exceed the Scheffé corrected F and hence the decision would be not to reject H_0. This can be compared with the decision to reject H_0 in the *a priori* uncorrected test. (Note that *a posteriori* tests are not available for within-subjects experiments.)

8.3.3 Newman–Keuls test

This test is appropriate for *a posteriori* comparisons of two conditions from among the set of conditions that make up a factor. A full discussion of it can be found in Winer *et al.* (1991).

The procedure differs from that for the Scheffé test. The *a posteriori* test of the comparison of **None** with **Partial** which now follows serves to illustrate the steps of the Newman–Keuls procedure.

Stage 1: Set out the means for all conditions in order of value.

38.75	47.00	60.63
None	**Complete**	**Partial**

Stage 2: Note the number of steps apart of the two means for which an *a posteriori* comparison is required (adjacent = 1 step, etc.).

None to **Partial** = 2 steps

The number of steps plus one is needed in using the tables of Studentized range in Appendix F. There, r represents the number of steps plus one (see Stage 4).

Stage 3: Calculate

$$q = \frac{\text{Mean}_1 - \text{Mean}_2}{(MS_{\text{error}}/n)^{1/2}}$$

where n is the number of measurements at each condition, Mean_1 and Mean_2 are the two means and MS_{error}, together with its df, is taken from the general analysis of variance F-test. Here,

$$q = \frac{60.63 - 38.75}{(216.42/8)^{1/2}} = 4.211$$

q is the test statistic whose value must exceed a critical q (represented as q_c) from the Studentized range table in Appendix F in order that H_0 can be rejected.

Stage 4: Compare with critical q from tables ($r=3$, $df=21$):

$q_c = 3.58$ at 0.05 significance level
4.64 at 0.01 significance level

Hence we can reject H_0 at the 0.05 level of significance as an *a posteriori* test.

Note: since direction is always known in *a posteriori* tests, no further adjustment is required for a directional test. Where it is appropriate, a directional test is assumed. Note also that *a posteriori* tests are not available for within-subjects experiments.

8.4 OVERVIEW OF DECISIONS FOR CONTRASTS AND COMPARISONS OF MEANS

Figure 8.2 is a **decision chart**. The researcher should start with the box labelled 'type of comparison' at the top. This is where the decision is made between *a priori* and *a posteriori*. For the *a posteriori* route the next

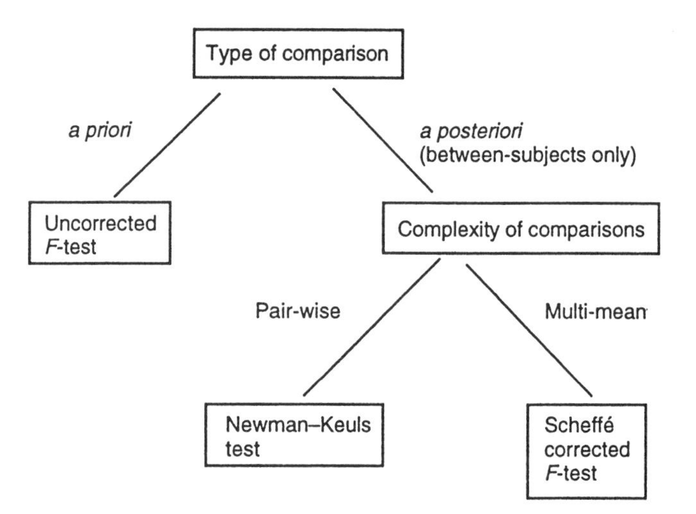

Fig. 8.2 Overview of decisions for contrasts and comparisons of means.

decision is pair-wise versus multi-mean. Multi-mean refers to a comparison or trend involving three or more means.

8.5 EXERCISES

8.1 The number of words recalled from a text was recorded for subjects who had spent various lengths of time studying it. One hundred and seventy five subjects were randomly allocated to five different study times resulting in five groups of 35 subjects.

The study times were 5, 10, 15, 20 and 25 minutes.

The mean numbers of words recalled were 9.2, 16.8, 24.0, 24.5 and 21.4, respectively.

The sum of squares between the groups was 5660.5, and within the groups was 110 939.

(a) Carry out an *a priori* test for a trend in the means of the groups in the direction of greater numbers of words recalled the longer the study time.
(b) Test the comparison, *a priori*, of the 5-minute group with the mean of all other groups.
(c) Test the comparison, *a posteriori*, of the means of the groups with study time 10 minutes or less with those groups with study times 15 minutes or more.
(d) Test the hypothesis, *a posteriori*, that the mean number of words recalled for the 5-minute groups is less than the mean for the 25-minute group.

9 Power and sensitivity in design decisions

9.1 INTRODUCTION

In an randomized experiment some of the variation in the dependent variable may be explainable by a variable other than those that are the main focus of attention. Such a variable is usually known as a covariate. A covariate can be incorporated into an analysis to improve the efficiency of the design. This may mean that the same probability of obtaining a significant result can be achieved with fewer subjects.

The cost of obtaining the covariate measurements may be prohibitively high. An example of a situation where this may be so is where the proposed covariate is an IQ score based on a 45-minute individual test session. The analysis technique was introduced and covered in detail for the continuous covariate in section 2.4 and Chapter 7, respectively. The category covariate was introduced in the context of the two-factor design in section 2.3.2. The two-factor design was dealt with in detail in Chapter 6.

Decision-making for the case of a continuous covariate is dealt with in section 9.2 and the parallel situation of the category-type covariate considered for use as a blocking factor is dealt with in section 9.3.

The choice of sample size is discussed for the single factor between-subjects design in section 9.4.

Finally, the choice of within-subjects or between-subjects design is dealt with in section 9.5.

9.2 SENSITIVITY AND EFFICIENCY GAINS FROM A CONTINUOUS COVARIATE

The example of the drugs experiment on rats introduced in section 2.4 and continued in Chapter 7 is used here to illustrate the decision of whether or not to use a continuous covariate in a single factor between-subjects randomized experiment. Two versions are compared: one with, and one without, the covariate.

The sensitivity formula is given in section 3.9. It is:

$$\text{sensitivity} = \frac{n}{MS}$$

Here n is the number of subjects experiencing each condition (or, more generally, the number of measurements obtained under each condition) and MS is the within-groups mean square (or other estimate of the between-subjects variance). This gives sensitivity values as set out in Table 9.1.

Since the numbers of subjects per group were identical in the two versions of the analysis, the sensitivities reflect the variance estimates. Thus the adjusted analysis is 0.0210/0.0091 = 2.308 times as sensitive as the unadjusted analysis.

Table 9.1 Sensitivity in the two versions

	n	*MS*	*Sensitivity*
Unadjusted	9	991	0.0091
Adjusted	9	429.5	0.0210

The comparison of sensitivity is useful if the covariate measurements are freely available to the experimenter. However, if account needs to be taken of the cost of obtaining the covariate measurements, the comparison of efficiencies is required.

To illustrate the calculation of efficiency, assumptions need to be made about the costs of setting up and running the experiment. The costs likely to be encountered were identified in section 3.10. The same cost headings are reproduced in Table 9.2 with reasonable but fictitious amounts of time in hours for each version of the experiment.

Table 9.2 Costs of the rats' pulse rates experiment

		Unadjusted	*Adjusted*
Cost of finding subjects	18 @ 0.5	9	9
Cost of setting up conditions		15	15
Cost of taking subjects through the conditions	18 @ 1.0	18	18
Cost of obtaining covariate scores	18 @ 0.25	0	4.5
Total cost (hours)		42	46.5

The efficiency is calculated using the formula presented in section 3.10 and reproduced here.

$$\text{efficiency} = \frac{\text{sensitivity}}{\text{cost}}$$

This leads to the values for efficiency shown in Table 9.3.

Table 9.3 Sensitivity and efficiency in the two versions

	n	*MS*	*Sensitivity*	*Cost*	*Efficiency*
Unadjusted	9	991	0.0091	42	0.000217
Adjusted	9	429.5	0.0210	46.5	0.000452

The comparison of alternative designs is best carried out in terms of their **relative efficiency** or R.E.

$$\text{Relative efficiency} = \frac{\text{efficiency of adjusted design}}{\text{efficiency of unadjusted design}} = \frac{0.000452}{0.000217}$$

$$= 2.083$$

This means that the adjusted design is more than twice as efficient as the unadjusted design.

The relative sensitivity of 2.308, can be compared to the relative efficiency of 2.083. The benefit of the adjusted relative to the unadjusted design is more marked in terms of relative sensitivity than in terms of relative efficiency. This is due to having taken account of the cost of the covariate in the efficiency calculation.

Simple pilot study and rule of thumb

In the event that a very simple rule of thumb is required, this would be expressed in terms of the size of the relationship of the covariate to the dependent variable. If at least 30% of the variation in the dependent variable is explained by the proposed covariate then it should be included in the experiment.

A simple pilot study would make possible the calculation of the correlation (r) of the covariate with the dependent variable. If r^2 exceeds 0.30 (i.e. if r exceeds 0.55) the rule of thumb would indicate a decision to include the covariate. Note that r^2 is the proportion of SS of the d.v. explained by the covariate.

In the rats experiment discussed above, the **weights** of the rats explain 9070/16 357 or 55% of the variation in the **pulse** rates. This is a clear indication under the rule of thumb that the **weights** should be made use of in the experiment.

9.3 SENSITIVITY AND EFFICIENCY GAINS FROM A CATEGORY COVARIATE

An example is used to illustrate the making of the decision to use or not use a category-type covariate as a blocking factor in a single-factor between-subjects randomized experiment. Reference could usefully be made to the discussion of the randomized block design in section 2.3.2.

9.3.1 Example

A researcher was designing an experiment to compare the effectiveness of three instruction techniques. This would require a single-factor between-subjects design with random allocation of subjects to conditions.

A simple pass/fail ability test was available which was under consideration as a way of grouping subjects according to ability. This would be used as a blocking factor. If it were used, the design would become a randomized block with a sample of **pass** subjects randomly allocated to the three instruction techniques and a parallel sample of **fail** subjects likewise randomly allocated to the three techniques.

Figure 9.1 represents the layout of the randomized block design. The dependent variable is a score on a test of **recall** of the material taught.

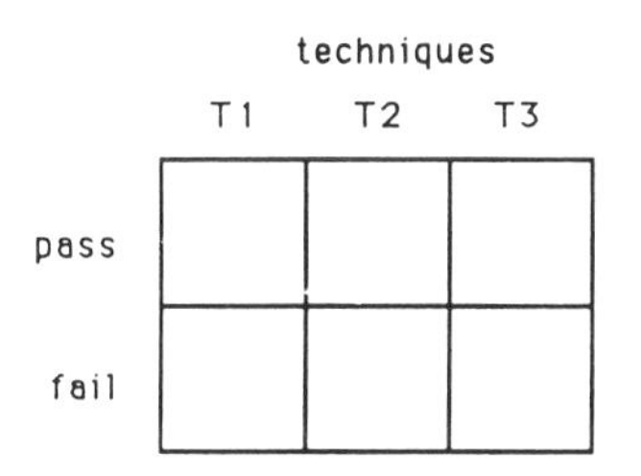

Fig. 9.1 Layout diagram for single-factor design with blocking.

There is a time cost associated with the use of the ability test. It takes 15 minutes per person tested. A decision has to be made on whether it is worth the time cost involved in using the ability test. A simple pilot study will provide enough information to enable the rule of thumb to be applied. A full pilot study will provide enough information to enable relative efficiency to be considered. Both approaches are illustrated here.

The simple pilot study estimated the strength of the effect of passing or **failing** the test on the individuals' **scores** on the dependent variable.

Simple pilot study and rule of thumb

If at least 30% of the variation in the dependent variable is explained by the proposed blocking factor then it should be included in the experiment.

The simple pilot study consisted of exposing 10 each of **pass** and **fail** subjects to one of the training conditions and then administering the **recall** test. The subjects who had **passed** the ability test obtained higher scores on the **recall** test. The analysis of variance summary table is set out in Table 9.4. This indicates that approximately 19% (i.e. 820/4320) of the total variation in scores is explained by **pass** or **fail** on the ability test. As a rule of thumb, if no other information were available, 30% could be regarded as the level at which a decision would be made to include **ability** in the design of the experiment. By this rule the ability test would not be used.

Table 9.4 Analysis of variance of simple pilot study

Source	*df*	*SS*
Ability	1	820
Within-groups	18	3500
Total	19	4320

Full pilot study and analysis of sensitivity and efficiency

By running a full pilot study which is a small scale version of the complete study it is possible to obtain information that leads to a decision about inclusion of the blocking factor in the complete study on the basis of relative efficiency.

The full pilot study was run with six subjects allocated to each combination of training technique in each ability group. This meant that 36 subjects were involved in all. The resulting analysis of variance summary table is set out in Table 9.5

Table 9.5 Analysis of the variance of the full pilot study

Source	*df*	*SS*	*MS*
Techniques	2	618	
Ability	1	889	
Interaction	2	490	
Within-groups	30	2323	77.4
Total	35	4320	

If **ability** were to be ignored, the *SS* for ability and for interaction and the associated degrees of freedom would be pooled into the within-groups terms to give the summary in Table 9.6.

Table 9.6 ANOVA of full pilot study ignoring blocking factor

Source	*df*	*SS*	*MS*
Techniques	2	618	
Within-groups	33	3702	112.2
Total	35	4320	

Comparison of sensitivities

The comparison of sensitivities amounts to a comparison of 12/112.2 with 12/77.4, i.e. 0.1069 compared to 0.1550. Sensitivity is 1.45 times higher in the version which includes the blocking factor. If there were no costs associated with the use of the blocking factor the higher sensitivity would be decisive. However, in this case there is 15 minutes per ability test to be taken into account.

Comparison of efficiencies

The comparison of efficiencies requires information on costs of setting up the research, obtaining subjects and carrying out all tests and measurements. Suppose costs were as set out in Table 9.7. The efficiency is

Table 9.7 Costs in hours

		Unadjusted	*Adjusted*
Cost of finding subjects	36 @ 0.25	9	9
Cost of setting up experiment		10	10
Cost of taking subjects through the conditions and completing d.v. tests	36 @ 1.25	45	45
Cost of obtaining covariate scores	36 @ 0.25	0	9
Total cost (hours)		64	73

calculated using the formula presented in section 3.10 and reproduced here:

$$\text{efficiency} = \frac{\text{sensitivity}}{\text{cost}}$$

This gives efficiencies as 0.1069/64 and 0.1550/73 for the versions with and without the covariate. The results are summarized in Table 9.8.

Table 9.8 Summary of sensitivity and efficiency

	n	*MS*	*Sensitivity*	*Cost*	*Efficiency*
Not blocked	12	112.2	0.1069	64	0.001671
Blocked	12	77.4	0.1550	73	0.002123

The comparison of alternative designs is best carried out in terms of their relative efficiency or R.E.:

$$\text{Relative efficiency} = \frac{\text{efficiency of blocked design}}{\text{efficiency of standard design}} = \frac{0.0021238}{0.0016711}$$

$$= 1.27$$

The relative sensitivity of 1.45 has become a relative efficiency of 1.27. This moderation of the benefit of the blocked relative to the not-blocked design is due to inclusion of the cost of the covariate in the efficiency calculation. This represents a clear decision to include the blocking factor. This contrasts with the decision obtained from the *rule of thumb* approach.

9.4 CHOICE OF SAMPLE SIZE

9.4.1 Introduction

Choice of sample size is often the most difficult part of the planning of an experiment. The feasibility of a project within certain time and cost constraints depends on the number of subjects required. The discussion is most usefully conducted in terms of power.

Power was defined in section 3.8 as the probability that in running the experiment, the experimenter will not make a type II error. This is the error of deciding not to reject a false null hypothesis. Power is increased when the sample size is increased or when the heterogeneity of the sample is effectively reduced either through grouping or blocking of subjects (section 9.3) or through adjustment based on a covariate (section 9.2).

If no covariates are available then the aim is to use a sample size which is large enough to provide adequate power but not unnecessarily large. It will be seen that increasing the sample size beyond a certain optimal level leads to minimal increase in power and is uneconomic.

9.4.2 Approximate sample size determination

The discussions on variance of means in section 3.2 and on confidence intervals in section 3.4 provide a simple way of looking at sample size. There are two requirements:

1. An estimate of the standard deviation or variance of the intended dependent variable in the population.
2. A view as to the smallest difference among the means that is of practical importance. The experiment will be designed to 'detect' differences of this size or larger.

If the smallest practical difference between a pair of means is represented by **spd** (the smallest practical difference between any two of the three groups whose comparison concerns the experimenter) and if **variance** stands for the variance of the dependent variable in a sample of subjects from the population, then n, the minimum number of subjects per group, is given by the formula:

$$\begin{aligned} n &= (2)^2(1.96)^2(\text{variance})/\mathbf{spd}^2 \\ &= (15.3664)\ (\text{variance})/\mathbf{spd}^2 \end{aligned}$$

(Note that the variance is equivalent to the standard deviation squared, $(\text{s.d.})^2$, and to the within-groups mean square.)

This formula can be relied on to give a power above 0.70 if at least five subjects are used in each condition being compared and if the 0.05 significance level is used for the test.

The rationale for this approximation is set out in Appendix E. Intuitively it can be justified as follows. The approximate 95% confidence interval about a sample mean is $\pm(1.96)(\text{variance}/n)^{1/2}$. From this it follows that $(2)(1.96)(\text{variance}/n)^{1/2}$ is the distance apart of means whose confidence intervals touch without overlapping. The approximate method chooses a sample size which leads to confidence intervals small enough to touch without overlapping when the means differ by the **spd**.

It is evident from this formula that n varies directly with the variance and inversely with the square of **spd**. This means that the smaller the variance (i.e. the more homogeneous the population or the less the random error in the measurements) the smaller the number of subjects required. Also, the larger the differences among the means that are regarded as worthwhile the smaller is the number of subjects required.

Numerical example

Consider the gerbil experiment in section 4.3.1. Suppose **spd** was thought to be 10 and the variance, based on measurements of the dependent variable on a sample of gerbils, was 200 (we know that the estimate from the ANOVA summary table in section 4.4.2 is 216). This leads to:

$$n = (15.3664)(200)/100$$
$$= 30 \text{ (approximately)}$$

This implies that approximately 90 gerbils would be needed, 30 per group.

9.4.3 Exact sample size and power determination

Introduction

A more accurate approach is available using tables of the **non-central** F. Much the same information is required. The procedure is best described through the use of an example.

Numerical example for power and sample size

An experiment is to be run with one **control** and two **interference** conditions. The dependent variable is **response time** in milliseconds. The **response times** are expected to be about 9 milliseconds longer in the two **interference conditions**. The standard deviation is known to be about 9.5 milliseconds in the **control** condition.

Step 1: Express smallest practical differences among means as deviations.

The means under influence of the conditions to be included in the experiment need to be expressed as deviations (section 2.1). Here the required deviations are (−6, +3, +3), corresponding to the **control** and two **interference** conditions, respectively, in order to accommodate the 9 millisecond difference. The corresponding bar chart is shown in Fig. 9.2.

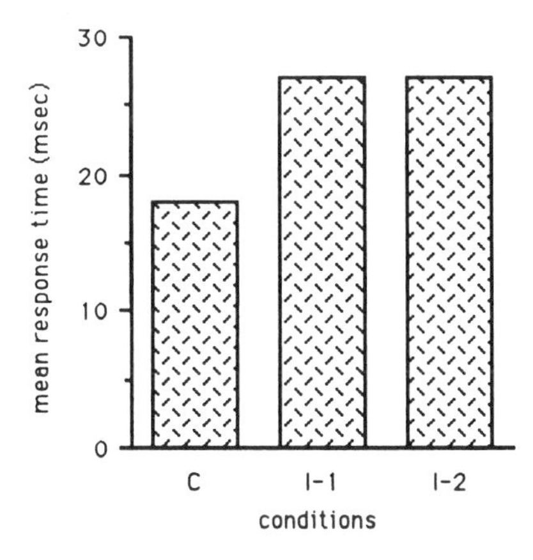

Fig. 9.2 Mean response times for a control (C) and two interference (I-1, I-2) conditions. Arbitrary overall mean = 24.

Step 2: Obtain estimate of variance.

This can be calculated directly or from the square of the standard deviation from a sample or, using previous results, may be taken as the mean square within-groups from an ANOVA.

Here we take

$$\begin{aligned}\text{variance} &= (\text{standard deviation})^2\\ &= 9.5^2\\ &= 90.25\end{aligned}$$

Step 3: Calculate ϕ.

Do this for a range of values of n, the number of subjects in any one condition. The formula for ϕ is

$$\phi = \left(\frac{n(t_1^2 + t_2^2 + t_3^2 + \cdots)}{(k)(\text{variance})}\right)^{1/2}$$

where k is the number of conditions in the experiment and t_1, t_2, etc. are the values of the deviations among the means.

Take values of n as follows: 4, 6, 10, 15, 20, 25, 30, 40, 50. For $n=4$:

$$\phi = \left(\frac{4(6^2 + 3^2 + 3^2)}{(3)(90.25)}\right)^{1/2}$$

$$= 0.893$$

The complete table of values of ϕ is set out, with power values for significance level 0.05, in Table 9.9.

Table 9.9 Values of ϕ and power for given values of n

Group size (n)	ϕ	*Power*
4	0.893	0.20
6	1.09	0.31
10	1.41	0.53
15	1.73	0.76
20	2.00	0.88
25	2.23	0.94
30	2.45	0.98
40	2.82	0.995
50	3.16	1.000

Step 4: Look up the values of power.

This is done in the tables of non-central F in Appendix F. In the tables of non-central F, (1 − power) is tabulated according to:

ϕ	the non-centrality parameter
$(k-1)$	the degrees of freedom between-groups
$k(n-1)$	the degrees of freedom within-groups (n is the number of subjects per group)
0.05 or 0.01	the significance level of the test

The table value is the probability of a type II error rather than power. Power is obtained as (1 − table value).

Continuing the example of $n=4$,

$$\begin{aligned} \phi &= 0.893 \\ (k-1) &= 2 \\ k(n-1) &= 9 \end{aligned}$$

This gives 0.80 and 0.94 for significance levels 0.05 and 0.01, respectively. (Note: interpolation was used within the table.) Taking these table values away from 1 gives power values of 0.20 and 0.06 respectively.

Step 5: Plot a graph.

The graph should be drawn with sample size on the horizontal axis and power on the vertical axis. Smooth the curve appropriately. This is shown in Fig. 9.3. It enables the researcher to read off the power for any value of the sample size and for either 0.05 or 0.01 significance level.

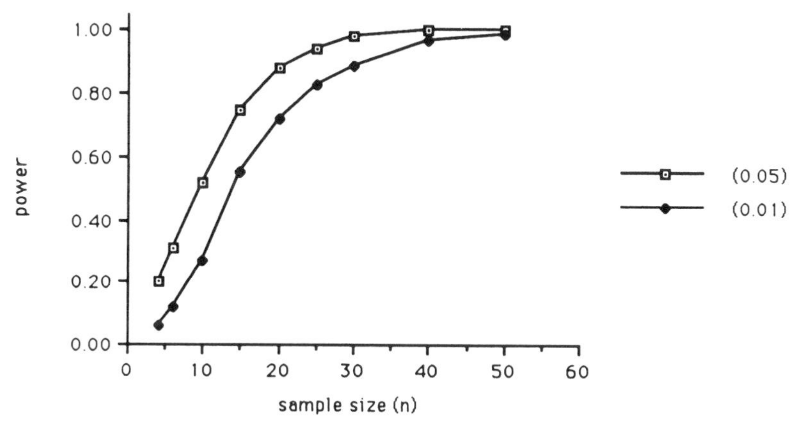

Fig. 9.3 Power versus sample size for significance levels 0.05 and 0.01.

Interpretation of sample size and power graph

The gradual flattening of the graph as sample size increases indicates the diminishing returns referred to above. A power of at least 0.7 is usually required. However, the individual circumstances will involve a variety of factors contributing to the final decision. For this reason it is worthwhile plotting the full graph.

9.5 CHOICE OF WITHIN- OR BETWEEN-SUBJECTS DESIGN

9.5.1 Introduction

This is based on the relative magnitude of the reliability *MS* and between-subjects *MS*, together with the relative costs of finding and re-testing sample subjects. The within-subjects or repeated measures design is almost always more efficient than the between-subjects or independent groups design.

9.5.2 Numerical example of relative efficiency of between- and within-subjects design

Consider the dowel balancing experiment described in section 5.4.1 The analysis of variance summary table (Table 5.4) shows the reliability and between-subjects *MS*s to be 4 and 20 respectively.

This implies that if four different subjects were used for each of the three conditions instead of the same four used repeatedly, the sensitivities would be 4/4 and 4/20, respectively. This indicates that the repeated measures design is 5.0 times as sensitive as the independent groups version.

Since obtaining new samples of subjects is almost always more expensive than having the original sample take part several times and since the reliability *MS* is almost certain to be smaller than the between-subjects *MS*, the usual rule is to use the repeated measures design whenever possible.

The main circumstance in which a new sample of subjects for each condition could lead to better efficiency would be if subjects needed a long rest between taking part in each condition.

9.5.3 Characteristics prohibiting use of within-subjects design

Introduction

Various restrictions apply to whether a factor can be within-subjects. The main points are summarized here:

1. Factors whose levels represent characteristics of the subjects which endure beyond the time-scale of the research (such as sex) are known as **intrinsic**. Intrinsic factors must be between-subjects.
2. Factors whose levels represent extrinsic conditions but which have a long-term effect on the subjects must be between-subjects. An example is where the conditions are alternative methods of teaching a given body of material. The effect of learning the material is long-term.
3. The process of measuring the response or dependent variable may have a long-term effect on the individual subject. This means the measurement can only be made once and implies use of a between-subjects design. An example would be a test of performance on which individuals improve markedly with practice.

If none of the above restrictions applies and no problems arise due to subject tiredness or boredom or other confounding effects, then the within-subjects factor is likely to be the most appropriate design.

9.5.4 Threats to validity of within-subjects designs

Validity is the extent to which the observed effect on the dependent variable is caused only by the values of the independent variable that define the conditions to be compared. Points that need to be taken into account are:

(a) Maturation or development of the subjects over the period of the experiment.
(b) Practice effect with the measuring instrument and increased familiarity with the situation.
(c) One condition influencing another (carry-over effect).
(d) Systematic drop-out (certain types of individuals inclined to drop out at certain stages).
(e) Regression to the mean. If individuals are selected according to their levels on a variable which is related (but not perfectly) to the dependent variable then at successive measurements the scores will be closer to the mean of the population from which the individuals were selected.

Most of the problems (a) to (e) can be reduced by randomizing the order in which subjects are exposed to the various conditions.

9.6 SUMMARY OF INFLUENCES ON DESIGN DECISIONS

(a) *Between-subjects with versus without continuous covariate*
Favour covariate if:
- Correlation of covariate with dependent variable is large.
- Cost of covariate measure is low relative to other costs.

(b) *Between-subjects with versus without category-type blocking factor*
Favour blocking factor if:
- Effect of blocking factor on dependent variable is large.
- Cost of grouping subjects by categories of the blocking factor is small relative to other costs.

(c) *Continuous covariate versus category covariate*
Favour continuous covariate if:
- Correlation of covariate with dependent variable is above 0.55.
- Inconvenient to group subjects before running experiment.
- Conditions × blocks interaction not of interest.
- ANCOVA analysis interpretation not feared for its complexity of interpretation.
- Assumptions for ANCOVA model are satisfied.

(d) *Within- versus between-subjects*
Favour within-subjects if:
- No logical or logistic objection to subjects being used repeatedly.
- Reliability *MS* is small relative to the between-subjects *MS*.
- Cost of obtaining sample subjects is high relative to the cost of single measurement on dependent variable.
- Number of conditions to be compared is large.

9.7 EXERCISES

9.1 Sixteen randomly selected subjects are allocated at random, four to each of four conditions: **counting**, **rhyming**, **adjective** and **imagery**. The dependent variable is the number of words recalled from a list. The age of each subject is recorded and is used as a continuous covariate with the aim of increasing the power of the test of significance of **conditions** in the ANOVA.

The *SS*s unadjusted and adjusted for **age** are in the following table:

Source	*SS*	SS_{adj}
Conditions	101	98
(Covariate	—	47)
Residual	154	110
Total	255	

Assume that it takes 0.25 hours to take each subject through the experiment, 0.5 hours to find each subject and 4 hours set-up time. Assume further that an additional 3 hours work is involved in carrying out an ANCOVA analysis which makes use of the subjects' ages additional to the work that would be involved in the ordinary ANOVA which ignores ages.

(a) Apply the three methods for comparing the versions of the design and analysis with and without use of the continuous covariate **age**:

A: rule of thumb
B: compare sensitivities
C: compare efficiencies

(b) Which of the three methods in (a) gives the strongest support for the use of **age** as a covariate?

9.2 A researcher who was designing an experiment to compare the effectiveness of three instruction **techniques** could not decide how to make use of the subjects' **intelligence** scores, which were available at no extra cost and which were correlated with the scores on the dependent variable. The dependent variable was a score on a test of recall of the material taught.

The researcher conducted the experiment as a 2 × 3 randomized block design with **intelligence** as a two-level blocking factor (higher, lower) and with four subjects per cell.

Because of indecision the experimenter carried out two analyses. The first analysis was the appropriate one for the randomized block:

Source	*df*	*SS*
Techniques	2	618
Blocks	1	889
Interaction	2	270
Residual	18	4044
Total	23	5821

The second analysis ignored the blocking factor and assumed a one-factor design with **intelligence** as a continuous covariate. The following table shows the analysis of variance for **techniques** adjusted for the effect of **intelligence**:

Source	*df*	SS_{adj}
Techniques	2	590
Residual	20	1498
Total	22	

(a) Complete the tests of significance for **techniques** for each analysis.
(b) Calculate the relative efficiency of the two approaches making clear which is the more efficient.
(c) Give your view of how the researcher should have proceeded in this case and outline the relative advantages and disadvantages, in general, of the two approaches.

9.3 A hospital's clinical staff agreed to run a trial to evaluate the effectiveness of individual therapy on schizophrenia.

Subjects would be randomly allocated to either the therapy or the control group. The dependent variable would be the rating on a personality scale. The trial would run for a year. At the end of this time a difference, on average, of 20 points on the personality scale, between the two groups, was the minimum that could be considered clinically useful.

(a) If the mean square within groups is known to be about 200 and a significance level of 0.05 was required, obtain an approximate estimate of the required sample size using the formula in section 9.4.2.
(b) Investigate graphically the relationship between power and number of subjects used in the trial supposing a significance level of 0.01 was to be used.
(c) How many subjects are needed to give a power of 50% when a 0.01 significance level is used?

PART 2

Unbalanced, Non-Randomized and Survey Designs

Unbalanced and confounded designs

10.1 INTRODUCTION

Research can lead to conclusions about one variable causing an effect on another variable. The requirement for this is that the researcher varies the values of the one variable and in doing so obtains an associated effect in the other variable. This is the classical experimental research method. Part One was concerned with this approach.

A variable consisting of conditions that can be freely varied by an experimenter and experienced by an individual who is a subject of the research is known as **extrinsic**. Part One was mainly concerned with experiments designed to test the existence of a causal effect of a category extrinsic variable on a continuous dependent variable.

An example was the gerbil experiment reported in Chapter 4. In this the researcher determined the degree of interruption to feeding experienced by each gerbil and looked for an effect on the number of times the gerbil returned to the feeding site. The variation of the degree of interruption was achieved by random allocation of experimental conditions to gerbils.

The uncertainty introduced to the situation by the presence of individual differences between the gerbils, provided it is of a random nature, is dealt with by the hypothesis test strategy.

In Part Two consideration is given to research designs in situations in which the researcher lacks the control required for a randomized laboratory-style experiment. One reason may be because **intrinsic** variables are the focus of interest. Intrinsic variables are relatively permanent characteristics of individuals, such as their sex or the number of children they have borne. Another reason may be that random allocation is not easily implemented outside the laboratory. The pressures of life in commerce, industry, institutions and the home represent the main area of difficulty.

Random allocation of individuals to categories of an extrinsic variable is not always possible. The researcher may have to allocate the different conditions to pre-existing groups defined by, for example, attending the same health centre, being pupils of the same school class or working on the same production line. This results in what is called a **quasi-experiment**.

The quasi-experiment has interpretation difficulties due to the differences between conditions being confounded with differences between health centres, classes or production lines and due to group differences in the characteristics of the individuals. Section 10.3 deals with techniques for removing differences between these pre-existing groups by **adjustment**.

Equal numbers of individual subjects in each condition or combination of conditions will not always be possible in real-world research. This problem can arise in a randomized experiment if one of the factors is an **intrinsic blocking factor.** For example, in an experiment to compare treatments for pre-menstrual tension it may be required to allocate women at random to one of three treatments. Suppose **parity** (no children, one child or 2+ children) is to be used as a blocking factor. It may be difficult to arrange to have the same number of women in each group if there are fewer women with 2+ children willing to take part. Also, the pattern of drop-out may differ between the groups.

The same problem of unequal numbers of subjects in groups is an inevitable part of research in which there is no random allocation or intervention by the researcher. This is the area of **observational research** or **survey research**, in which individuals are identified or classified according to levels of the intrinsic independent variables and measured for a continuous score on the dependent variable. For example, consider a survey intended to study the effects of **smoker status** (non-smoker, ex-smoker, current smoker) and **family history** of circulation problems (yes, no) on **blood pressure**.

This is a two-factor independent groups design with **blood pressure** the continuous scaled dependent variable. It is analysed by analysis of variance. However, it is almost certain that there will be different numbers of individuals classified into each of the six cells according to their categories of the two independent variables. The analysis of variance technique described for the two-factor design in Chapter 6 will not work in this situation.

Unequal numbers leads to what is called an **unbalanced design** or **non-orthogonal design**. There are difficulties of interpretation as interdependencies are present between the 'independent' variables. Section 10.2 deals with this.

10.2 TWO-FACTOR UNBALANCED DESIGN

10.2.1 Introduction

Consider the example referred to in section 10.1 of a survey of systolic blood pressure (**BP**) of individuals classified according to smoking status (**status**) and family history of circulation and heart problems (**history**). This is a 2×3 independent groups design along the pattern of the machining example in Chapter 6. The one important difference is that the numbers of individuals in the cells are not controlled to be the same throughout. Rather, they reflect the frequencies of occurrence in the population of individuals with particular combinations of the intrinsic factors, **status** and **history**.

If this study had been contrived to have identical numbers in the cells, the marginal means for **status** would have described the effect of smoking on blood pressure without any confounding effect of family history. This confounding effect of **history** would have been removed in the design of the study. Unequal distributions of family history of circulation problems in the **smoker status** groups could not then serve as a possible alternative explanation of an observed relationship between smoking and blood pressure.

Analysis of the unbalanced design aims to remove confounding effects at the analysis stage that could, in principle, have been removed at the design stage.

The numbers of individuals in each cell and the mean **BP** are displayed in Table 10.1. The frequencies vary considerably from cell to cell. They range from 4 to 8; a factor of two. To continue sampling until there were enough subjects to provide 8 for each cell would probably require finding a further 38 subjects. Twenty-eight of these would have to be discarded. Most researchers prefer to use an analysis technique that copes with the unbalanced numbers.

The full data for the 38 individuals are presented in Table 10.2.

Table 10.1 Means of systolic **blood pressure**

		Smoker status		
		Non	**Ex**	**Current**
Family history	present	$n=7$ 126.57	$n=6$ 121.67	$n=7$ 131.57
	absent	8 111.75	4 108.25	6 128.17

Table 10.2 Blood pressure, family history and **smoker status**

	Smoker status		
	Non	**Ex**	**Current**
Family history present	125	114	135
	156	107	120
	103	134	123
	129	140	113
	110	120	165
	128	115	145
	135		120
Family history absent	114	110	140
	110	128	125
	91	105	123
	136	90	108
	105		113
	125		160
	103		
	110		

Two issues arise in connection with this type of design which lead to its exclusion from introductory courses in statistics. They are the estimation of the row and column marginal means and the tests of significance. These will be dealt with in sections 10.2.2–10.2.4.

10.2.2. Estimation of row and column marginal means

If the sampling of the blood pressure survey had been so arranged that there were equal numbers of subjects in the six cells then the marginal row and column means would be expected to have the values shown in Table 10.3. That is, they would be the means of the means in the corresponding row or column. These are called the **unweighted means**. The unweighted means are very easily calculated by hand. For example, in the first column, 119.16 is the mean of 126.57 and 111.75.

Table 10.3 The unweighted marginal means

		Smoker status			
		Non	**Ex**	**Current**	
Family history	**Yes**	7 126.57	6 121.67	7 131.57	126.60
	No	8 111.75	4 108.25	6 128.17	116.06
		119.16	114.96	129.87	

The column marginal means 119.16, 114.96 and 129.87 are the mean blood pressures for the three categories of **smoker status** adjusted for the effects of **family history**. Likewise 126.60 and 116.06 are the mean blood pressures for the two categories of **family history** adjusted for the effects of **smoker status**.

If **family history** was ignored, the mean **blood pressures** for the three **smoker status** categories, based respectively on 15, 10 and 13 individuals, would be calculated as 118.67, 116.30 and 130.00. These are the **weighted means**. The name refers to the fact that they are means weighted for the numbers in the cells. They can be obtained as averages of all blood pressures in the appropriate category or from the cell means weighted by the frequencies. The weighted mean of the first column is obtained from the cell means as follows:

$$\text{weighted mean} = \frac{(7)(126.57) + (8)(111.75)}{7+8} = 118.67$$

The weighted mean **blood pressures** for the categories of **family history** are 126.85 and 116.44 based on 20 and 18 individuals respectively. The same formula is illustrated for the first of these:

$$\text{weighted mean} = \frac{(7)(126.57) + (6)(121.67) + (7)(131.57)}{7+6+7} = 126.85$$

The weighted means could present a misleading picture if **family history** had a large effect on **blood pressure** and disproportionate numbers with positive **family history** were present in the three groups. The unweighted means are usually more appropriate. They are printed out under the names **unweighted means** or **adjusted means** by computer programs that provide the **unique** or **Type III sums of squares**. (The sums of squares are discussed in section 10.2.3.)

No problem arises for the interaction. This is because it is described by the individual cell means. No adjustment is required.

10.2.3 Sequential and unique sums of squares and Venn diagrams

Unequal numbers of individuals in the cells results in **interdependence** among the independent variables. This requires the calculation of special types of sums of squares which facilitate interpretation and significance tests.

The statistical computer package will usually offer two types of sums of squares. They are the **sequential sums of squares**, also known as **Type I sums of squares**, and the **unique sums of squares**, also known as **Type III sums of squares**. These are best understood with the aid of a Venn diagram.

A **Venn diagram** is a rectangle representing the total sum of squares of the dependent variable. Regions of the square are marked out with circles to indicate portions of the total sum of squares explained by certain independent variables. In other words, the Venn diagram shows how the 'cake' is divided up and how much remains unallocated.

Several important concepts related to sums of squares (*SS*s) explained by interdependent independent variables can be represented by the Venn diagram. They are:

Sequential
Synergy
Unique
Adjusted

Sequential (sequential sums of squares)

This describes an analysis approach in which the independent variables (i.v.s) are used to explain the *SS* of the dependent variable (d.v.) in a particular sequence.

Figure 10.1(a) illustrates this for the sequence A then B, where A and B stand for any two factors. A is given the first bite at the cake. It explains a portion of the *SS* of the d.v. (Fig. 10.1(a1)).

In Fig. 10.1(a2), factor B, a second i.v., is included in the model. Factor B has the second bite at the cake. B explains as much as possible of the *SS* of the d.v. not already explained by A.

Note that B is shown overlapping A. This overlap region represents a portion of the variance of the d.v. explainable by either A or B. This means that one i.v., in the presence of the other i.v., explains less than if it were alone in the model. Overlap means that together A and B explain less of the variation of the d.v. than the sum of what each can explain alone.

The amount of *SS* explained by a factor therefore depends on the sequence in which the factors are included in the model.

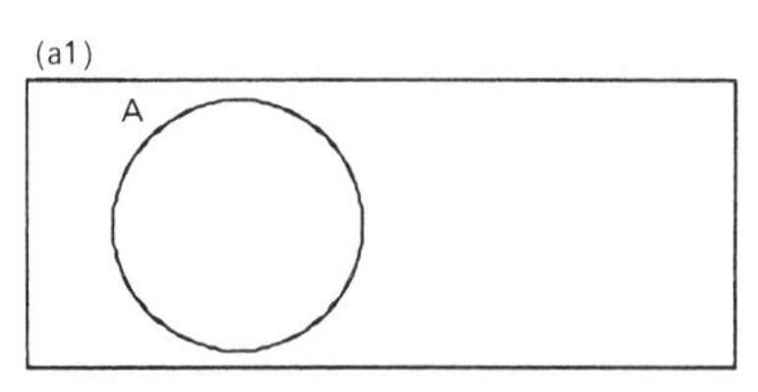

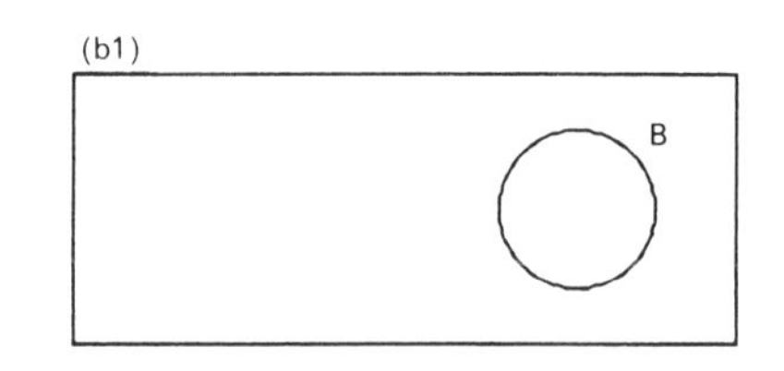

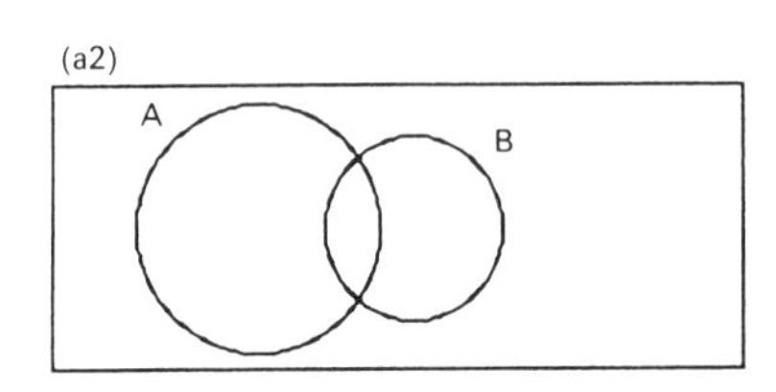

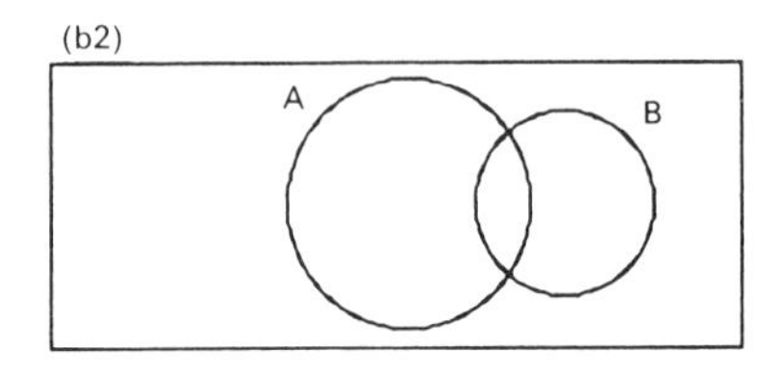

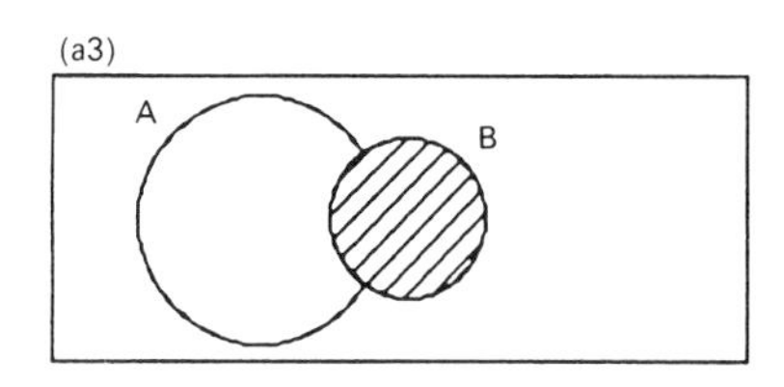

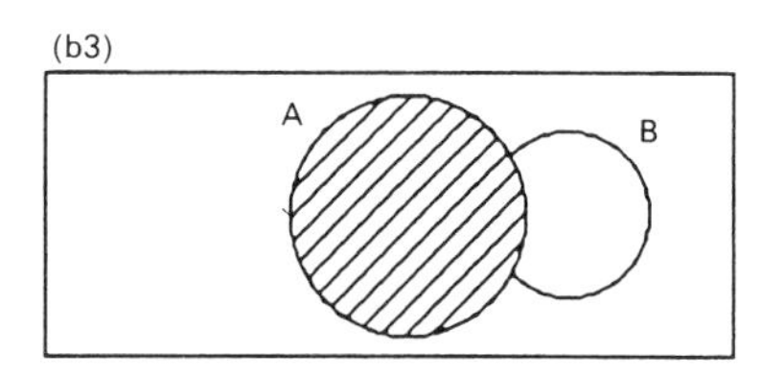

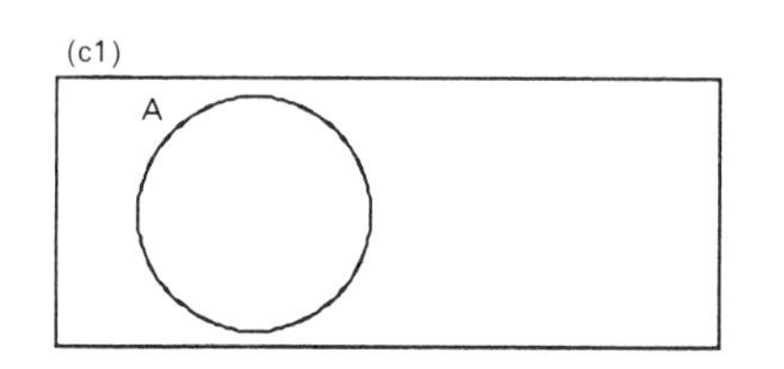

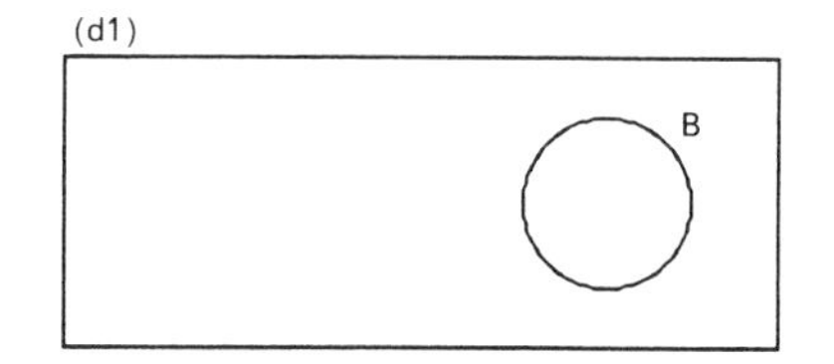

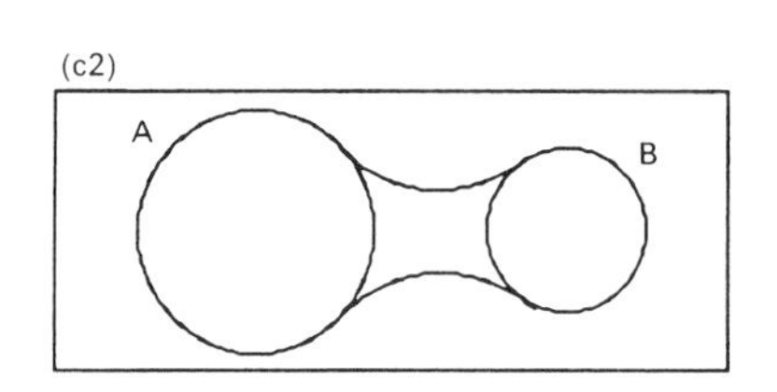

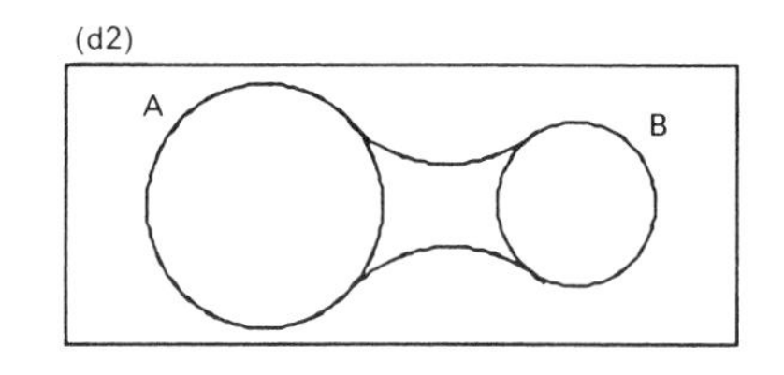

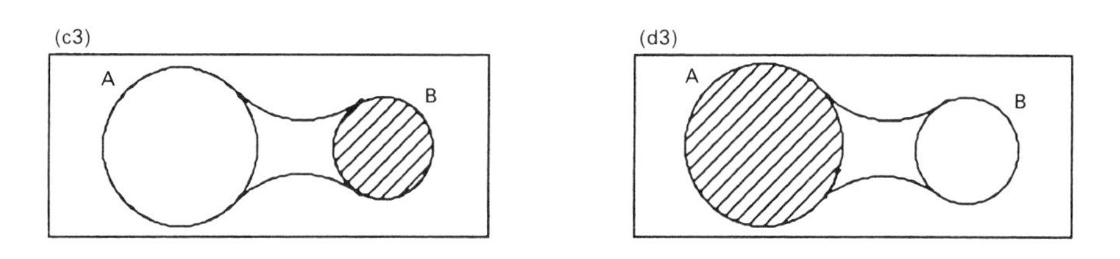

Fig. 10.1. Variance explained by interdependent variables A and B.

Suppose the order were reversed as in Fig. 10.1(b). This time B commences by explaining as much as possible. A follows, explaining as much as possible of what is not already explained by B.

Synergy (synergic sums of squares)

Sometimes interdependent factors A and B have a **synergic** relationship. This means that together they explain more of the variation of the d.v. than the sum of what each can explain alone. In this case the overlap on the Venn diagram is replaced by a concave region called the **synergy**. (The word 'synergy' is not normally used to refer to this concept, although its meaning exactly represents it.)

This is illustrated in Fig. 10.1(c). A enters first. The circular region labelled A represents the *SS* explained by A. B follows. The *SS* explained additionally by B is represented by the circular region labelled B together with the synergy.

Likewise, Fig. 10.1(d) represents the synergic situation with B starting and A following. This time A brings with it the synergy.

Unique (unique sums of squares)

This refers to the sum of squares of the d.v. explained by an independent variable when it is last in the sequence. In the overlapped diagrams in Figs 10.1(a2) and 10.1(b2) it is represented by the half-moon shaped parts of B and A respectively. In the synergy diagrams in Figs 10.1(c2) and 10.1(d2) it is represented by the circles for B and A, respectively, with synergy appended.

Adjusted (adjusted sums of squares)

The idea of adjustment was introduced in Chapter 7 and is further developed in section 10.3. The variation of the d.v. explained by factor A 'adjusted for the effect of factor B' refers to the variation that factor A would explain if both it and the d.v. were adjusted to the values they would have if factor B was fixed at a particular value (for all subjects).

On the Venn diagram this is represented by removing the region representing the 'partialled out' variable. In Fig. 10.1 the shaded region represents the partialled out i.v. When the shaded region is removed the total is reduced.

In Fig. 10.1(a3) the adjusted *SS* for A is less than the unadjusted *SS*. However, since the total is also reduced, the amount explained by A as a proportion of the adjusted total may be higher or lower.

In Fig. 10.1(c3) the adjusted *SS* for A is greater than the unadjusted *SS*. Since the total is again reduced, the amount explained by A as a proportion of the adjusted total is higher.

Figures 10.1(b3) and 10.1(d3) illustrate the equivalent situation for B adjusted for the effect of A.

(Note that 'effect of A with B partialled out' is equivalent to 'effect of A adjusted for the effect of B'.)

10.2.4 Tests of significance for the blood pressure example

Test of smoker status

Refer to the Venn diagram in Fig. 10.2(a). The Venn diagram was constructed from the results of two computer analyses for sequential or Type I sums of squares, one with the sequence **status** then **history** and one with the reverse order sequence. These provided the following results:

Source	*SS*	*Source*	*SS*
Status	1323.33	**History**	1025.71
History	1036.81	**Status**	1334.43
Interaction	252.78	Interaction	252.78
Residual	8473.84	Residual	8473.84

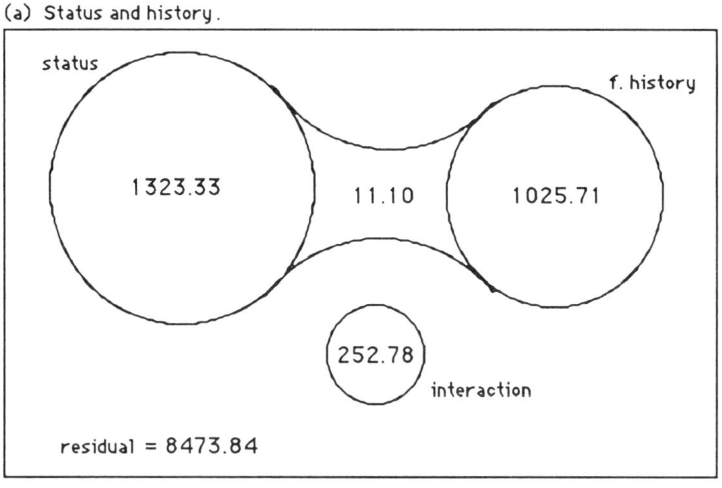

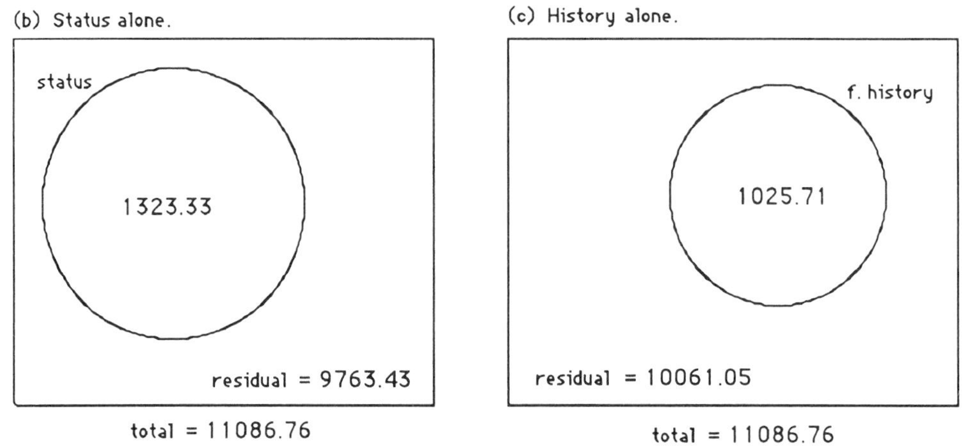

Fig. 10.2 Sums of squares of blood pressure explained by two variables.

Ignoring **history**, the test of **status** alone is based on the *SS*s in Fig. 10.2(b), which are set out formally in Table 10.4. The degrees of freedom are found from the general rules set out in sections 4.5 and 6.6.

Table 10.4 Test of **smoker status** ignoring **family history**

Source	*df*	*SS*	*MS*	*F*
Status	2	1323.33	661.665	2.37[a]
Residual	35	9763.43	278.955	
Total	37	11086.76		

[a]$F_c = 3.28$ at 0.05
$= 5.29$ at 0.01

This is the analysis that would have been obtained if no information was available on **history**. It is not recommended if **history** data is available.

A more useful analysis is obtained by partialling out or adjusting for the effect of **history**. This can best be understood with aid of the Venn diagram in Fig. 10.2(a). The adjusted summary table is obtained by erasing the regions corresponding to **history** and **interaction**. This results in the sums of squares in Table 10.5. This presents a test of the adjusted or unweighted means and corresponds to the result that might have been obtained if there had been equal numbers in the cells.

Table 10.5 Test of **smoker status** adjusting for **family history**

Source	df_{adj}	SS_{adj}	MS_{adj}	F_{adj}
Status	2	1334.43	667.215	2.52[a]
Residual	32	8473.84	264.808	
Total	34	9808.27		

[a]$F_c = 3.32$ at 0.05
$= 5.39$ at 0.01

Similar analyses can be carried out for the test of significance of **family history**. The test ignoring **smoker status** would use *SS*s 1025.71 and 10 061.05 with 1 and 36 degrees of freedom for **history** and **residual**, respectively, as shown in Fig. 10.2(c). The recommended test which adjusts for the effect of **smoker status** is in Table 10.6.

Note that the adjusted totals have different degrees of freedom in Tables 10.5 and 10.6. The unadjusted *df* for total is one less than the number of

Table 10.6 Test of **family history** adjusting for **smoker status**

Source	df_{adj}	SS_{adj}	MS_{adj}	F_{adj}
History	1	1036.81	1036.81	3.92[a]
Residual	32	8473.84	264.808	
Total	33	9510.65		

[a]$F_c = 4.17$ at 0.05
$= 7.56$ at 0.01

d.v. measurements in the analysis. The degrees of freedom lost are those corresponding to the regions erased from the Venn diagram in the adjustment procedure. Thus in Table 10.5, one and two degrees of freedom are lost for the erasure of **family history** and **interaction**, respectively. Likewise, in Table 10.6, two degrees of freedom are lost for both the erasure of **smoker status** and of **interaction**.

Test of significance for interaction

No adjustment is required for interaction. The test is constructed in the same way as for the balanced design. The summary is in Table 10.7. This table is interesting because it is free of any adjustment or need for adjustment. It is **orthogonal** in the sense that the total *SS* is partitioned into non-overlapping components. This property is possessed by balanced designs (i.e. designs with identical numbers of subjects in the cells). Also, it is interesting to note that the combined main effects reaches statistical significance at the 0.05 level ($p=0.046$).

Table 10.7 Test of interaction

Source	*df*	*SS*	*MS*	*F*
Main effects	3	2360.14	786.71	2.97[a]
Interaction	2	252.78	126.39	0.48[b]
Residual	32	8473.84	264.81	
Total	37	11086.76		

[a] $F_c=2.92$ at 0.05
[b] $F_c=3.32$ at 0.05

Thus the researcher could conclude, on the basis of this study, that the combined effect of **family history** and **smoking status** explains a significant portion of variation in systolic **blood pressure**. Note that 'explains variation in' is used rather than 'causes' as there has been no random allocation. Many possible confounding variables, such as age, have not been taken account of, either in the design or in the analysis.

10.2.5 Further example of the two-factor unbalanced design

To illustrate the procedures introduced above, an example follows whose sums of squares lead to an overlap of regions on the Venn diagram.

Two methods are to be compared for keeping pupils quiet during a private reading session. A **reward** method is used with one class and a **punishment** method with another, convenient, class in the same school. No random allocation was possible as for logistic reasons the pupils had to be taught in the same group at all times. This is a **quasi-experiment**.

The number of times each child made a **noise** during the experimental lesson was recorded and used as the dependent variable.

The scores and means for the two groups of children were:

Group	*Scores*	*Mean*
Reward	4 1 2 7 5 6 7 8 8 8 6 7 8 10	6.21
Punishment	3 2 4 3 1 4 6 3 8 8 4 8 5	4.54

The researcher was also interested in the effect of academic **ability** on the amount of **noise** made by pupils. The pupils had been previously categorized into ability bands **A**, **B** and **C**. This second factor was therefore included at the analysis stage.

The scores are presented in Table 10.8, classified according to **group** and **ability** and with cell means. It is evident that there are more high ability band pupils (band **A**) in the **punishment** group. **Ability** might be the explanation of the lower **noise** level (4.53 compared to 6.21) rather than the method of class control.

Table 10.8 Data and means for the pupil noise example

	Ability band					**Ability** band			
	A	B	C	mean		A	B	C	mean
Reward	4	7	8		**Punishment**	3	6	8	
	1	5	8			2	3	5	
	2	6	6			4	8		
		7	7			3	8		
		8	8			4	4		
			10			1			
Means	2.33	6.60	7.83	6.21	Means	2.83	5.80	6.50	4.53

The unweighted means are:

A	**B**	**C**	**Reward**	**Punishment**
2.58	6.20	7.17	5.58	5.04

It is evident from an examination of the adjusted (unweighted) means (5.58 and 5.04) that almost all the difference in method of class control has been removed by the adjustment made for **ability**.

Obtaining the sums of squares

The sums of squares can be obtained from the results of an analysis of variance from a statistical computer package. It is recommended that they be transferred to a Venn diagram in order to clarify the situation. Differences in procedure and vocabulary between statistical packages easily lead to misunderstandings. The user who has difficulty constructing the Venn diagram is at risk of making an error of interpretation of the analysis of variance summary table provided by the package. Usually Type I (sequential) analyses need to be run twice with the main effects in different sequences.

A single **adjusted** analysis sometimes makes it possible to omit the Venn diagram. Unfortunately, in the less good computer programs, it is not always clear what form of adjustment is being carried out. Only if Type III (or unique) is specified should the program be used.

The Venn diagram

The Venn diagram in Fig. 10.3 shows the large amount of overlap of the region representing **method**. This confirms that there is a large effect of adjustment on this variable. Ninety-two per cent of the variation explained by **method** is explainable by **ability** (17.43/18.93).

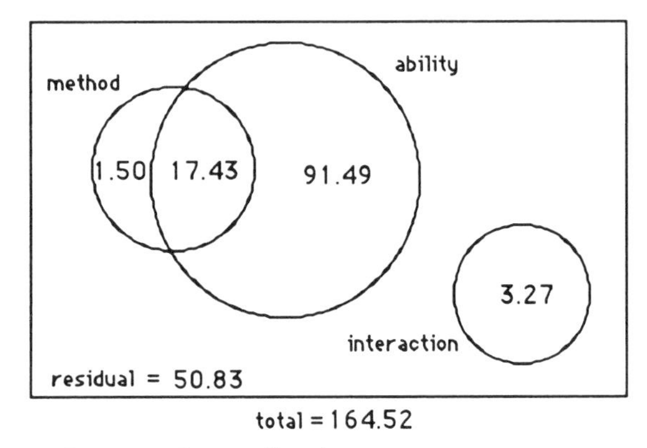

Fig. 10.3 Sums of squares for pupil noise.

The tests of significance

The three tests of significance for main effects and interaction follow in Tables 10.9 to 10.11.

Table 10.9 Test of **method** adjusted for **ability**

Source	df_{adj}	SS_{adj}	MS_{adj}	F_{adj}
Method	1	1.50	1.50	0.62[a]
Residual	21	50.83	2.42	
Total	22	52.33		

[a] $F_c = 4.32$ at 0.05

Table 10.10 Test of **ability** adjusted for **method**

Source	df_{adj}	SS_{adj}	MS_{adj}	F_{adj}
Ability	2	91.49	45.75	18.90[a]
Residual	21	50.83	2.420	
Total	23	142.32		

[a] $F_c = 5.79$ at 0.01

Table 10.11 Test of interaction

Source	*df*	*SS*	*MS*	*F*
Main effects	3	110.42	36.81	15.21[a]
Interaction	2	3.27	1.635	0.68[b]
Residual	21	50.83	2.420	
Total	26	164.52		

[a] $F_c = 4.88$ at 0.01
[b] $F_c = 3.49$ at 0.05

Conclusions

Only the proportion of the variation in classroom **noise** explained by **ability** is significant. However, other variables not included in the analysis, such as age and sex, could possibly explain the relationship of **ability** with **noise** found in this study.

10.3 CONFOUNDING IN ONE-VARIABLE NON-RANDOMIZED DESIGNS

10.3.1 The confounding problem

Compare the randomized and non-randomized designs whose aim is to research the effect on a continuous dependent variable **Z**, of a category independent variable **A** with two levels (**A**, **no A**).

Figure 10.4 illustrates the design and results of the randomized version.

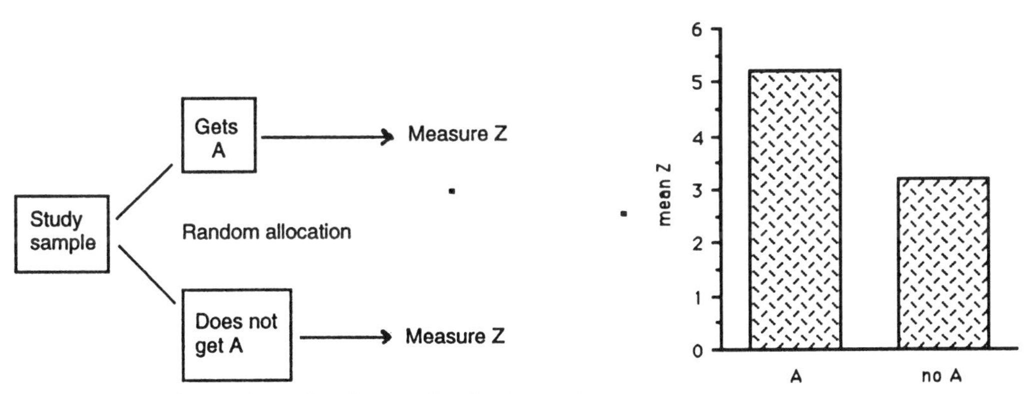

Fig. 10.4 Design and results of a randomized experiment.

If the means of **Z** in the two groups are significantly different, **A** is immediately established as having a **causal influence** on **Z**.

Suppose the true effect of **A** on **Z** is represented by deviations (+1, −1) for (**A**, **no A**). This is represented in the bar chart in Fig. 10.4.

Figure 10.5 illustrates the design and results of the non-randomized version, in which the samples differ in the proportions of individuals with the

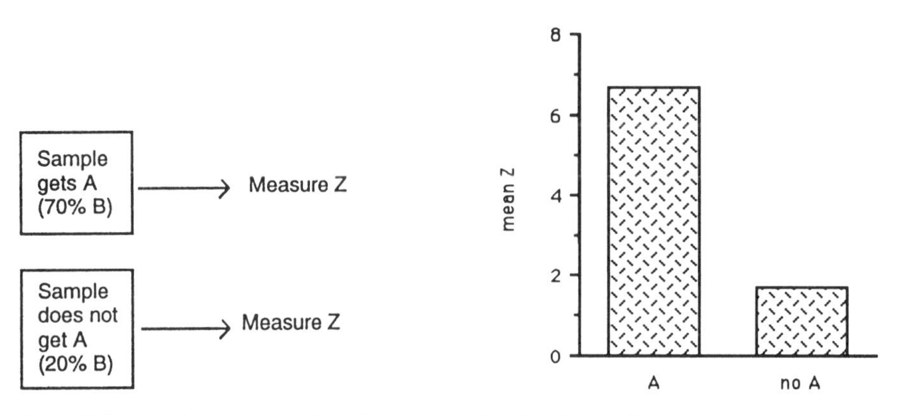

Fig. 10.5 Design and results of a non-randomized experiment.

characteristic **B**. Seventy per cent of members of the sample given condition **A** also have **B**, whereas only 20% of those not given **A** have **B**.

B influences the value of **Z**. The true effect of **B** on **Z** is represented by deviations (+3, −3). It is supposed there is no interaction between **A** and **B**.

The combined effect of **A** and **B**, using the unweighted means approach discussed in section 10.2.1, is shown in the bar chart in Fig. 10.5. The effect of **A** has been greatly enhanced by the contribution of **B**.

B, distributed unevenly across the groups, could create an apparent effect of **A** even if, in reality, there were none. **B** is acting as a **confounding variable**.

Even if the means in the two groups are significantly different, **A** cannot be said to have a causal influence on **Z** because the two groups are known to differ in the proportion of sample members with characteristic **B**.

B is a confounding variable if it influences **Z** and is distributed unevenly across the two groups.

The confounding variable **B** can be controlled in the analysis if subgroups with and without **B** are analysed separately. The result of this is shown in Fig. 10.6. Note that the effect of **A** is shown correctly as (+1, −1) in both bar charts.

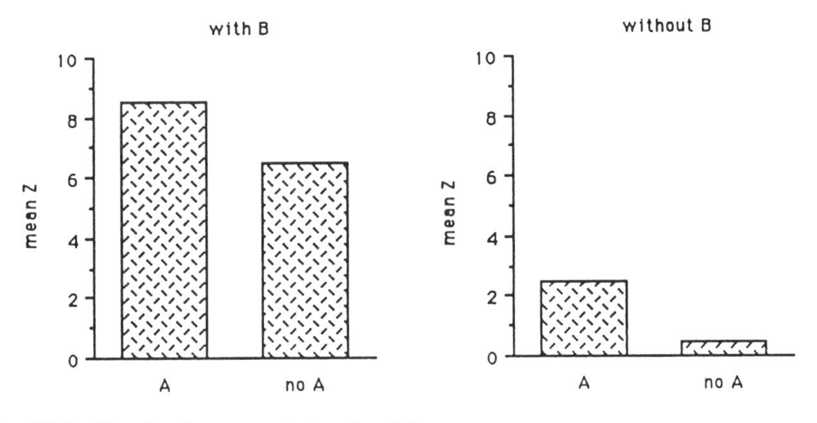

Fig. 10.6 Results for separate levels of B.

No simple statement of the effect of **A** on **Z** can be made until allowance or adjustment has been made for the effect of **B**. The problem, stated another way, is one of validity. The apparent effect of the independent variable on the dependent variable is **not valid**. Validity can be restored, to some degree, if the effect of the confounding variable can be removed.

Another way of expressing this is to say that the effect of the confounding variable has to be partialled out of the relationship of the independent to the dependent variable. Different techniques need to be employed depending on the type, category or continuous, of each of the three variables involved.

Definition of a confounding variable

A confounding variable is one whose effect on the dependent variable (d.v.) could be mistaken for the effect of the independent variable (i.v.). It is a variable which co-varies with both the i.v. and the d.v.

A variable which has no direct (or main) effect on the d.v. may influence the relationship of the i.v. to the d.v.; such a variable would interact with the i.v. The discussion in this section deals with confounding variables which have main, rather than interaction, effects on the d.v.

10.3.2 Overview of the confounding variable situations

All four designs listed in Table 10.12 could be **surveys**. The arrangements with a category-independent variable (numbers 1 and 2) are likely also to arise from a quasi-experiment. Different levels of the independent variable would be assigned to pre-existing groups of individuals and an effect would be sought on the dependent variable.

Table 10.12 Confounding variables in four types of surveys and quasi-experiments

Types of variables for main relationship		*Type of confounding variable*	*Analysis for hypothesis test and adjustment*
i.v.	*d.v.*		
1 Category	Continuous	Continuous	ANCOVA
2 Category	Continuous	Category	Two-factor ANOVA
3 Continuous	Continuous	Continuous	Partial correlation
4 Continuous	Continuous	Category	Role-reversed ANCOVA

There are four further designs corresponding to those in the table, but with a category-dependent variable. They are dealt with by the approach known as **log-linear modelling**. An accessible treatment is available in Knoke and Burke (1983).

Examples of the four arrangements of confounded designs (Table 10.12)

I.v. category, d.v. continuous – a between-subjects design

The Home Office has commissioned research on the effect of **type of conviction** on **locus of control** scores on prisoners. The two types of conviction to be compared are **violence** and **burglary**. **Locus of control** is to be measured using a scale of 1 to 7 on prisoners who have reached the end of their first four weeks of imprisonment. The **locus of control** scores are low for 'external' and high for 'internal' individuals.

1. Continuous confounding variable: **intelligence score** (IQ)
2. Category confounding variable: **sixth form education** (yes/no)

I.v. continuous, d.v. continuous – a correlation design
A teacher wishes to research the effect of **height** of pupils in a local junior school on peer-rated **popularity scores**. The measure of **popularity** will be on a scale from 0 to 20. Correlation will be used to analyse the data.

3. Continuous confounding variable: **age in months**
4. Category confounding variable: **sex** (male/female)

10.3.3 Adjustment for the between-subjects design

Introduction

Consider the example introduced in section 10.3.2 of a survey in which interest is centred on the effect of **type of conviction** on **locus of control** in recently convicted prisoners. **Intelligence** and **level of education** are identified as possible confounding variables because they are seen as likely to differ between the two groups of prisoners and to relate to the dependent variable, **locus of control**.

The data for 20 subjects are set out in Table 10.13. The **intelligence** scores follow the standard scale of 100 for population mean and are high for greater intelligence.

Continuous covariate

The analysis of covariance technique introduced in Chapter 7 is used here for adjusting the scores on the dependent variable to what would be expected if every prisoner had the same **IQ** score.

The summary tables showing the unadjusted and adjusted sums of squares are set out in Tables 10.14 and 10.15. Note that the sum of squares for **type of conviction** is greatly reduced by adjustment. This is because some of the variation in **locus of control** scores is explainable by the prisoners' **IQ** scores. The main consequence of this is that **conviction** changes from significant at the 0.05 level in the unadjusted analysis to not significant.

The mean **locus of control** scores in the two groups are changed by the adjustment procedure. This is expected since the mean **IQ** scores differ in the two groups. All mean scores are shown in Table 10.16.

The reduction in the magnitude of the effect of **type of conviction** on **locus of control** due to adjustment is seen in the mean scores as well as in the sums of squares. The adjusted difference between the means is only 0.46 compared to 0.76.

Table 10.13 Data for prisoners' **locus of control**

Conviction	**Locus of control**	**IQ**	**Sixth form**
Burglary	5.1	98	no
Burglary	3.4	84	yes
Burglary	3.9	79	yes
Burglary	4.7	72	no
Burglary	5.2	126	no
Burglary	4.6	84	no
Burglary	3.2	80	yes
Burglary	5.0	114	no
Burglary	4.4	108	yes
Burglary	4.5	89	no
Violence	4.0	90	no
Violence	3.5	92	no
Violence	3.1	81	yes
Violence	4.4	97	no
Violence	2.3	70	no
Violence	2.7	66	no
Violence	4.3	78	yes
Violence	3.8	85	no
Violence	4.4	103	no
Violence	3.9	76	no

Table 10.14 Unadjusted analysis of variance

Source	*df*	*SS*	*MS*	*F*
Conviction	1	2.89	2.89	5.698[a]
Residual	18	9.12	0.507	
Total	19	12.01		

[a] $F_c = 4.41$ at 0.05
$= 8.29$ at 0.01

Table 10.15 Adjusted analysis of variance (**IQ** removed from effect of **conviction**)

Source	df_{adj}	SS_{adj}	MS_{adj}	F_{adj}
Conviction	1	0.98	0.98	3.091[a]
Residual	17	5.40	0.318	
Total	18	6.38		

[a] $F_c = 4.45$ at 0.05
$= 8.40$ at 0.01

Table 10.16 Means for prisoners' **locus of control**

Group	*Mean* **IQ**	*Mean* **locus**	*Adjusted mean* **locus**
Burglary	93.4	4.40	4.25
Violence	83.8	3.64	3.79

The model fitted by the ANCOVA procedure as described in section 7.4 is:

Expected score for a randomly sampled subject	=	overall mean	+	conviction effect	+	covariate effect
	=	4.02	+	$\begin{Bmatrix} +0.23 \\ -0.23 \end{Bmatrix}$	+	0.0304 (covar − 88.6)

The deviations ± 0.23 are the amounts by which **locus** is increased or decreased from the overall mean due to **type of conviction**. 0.0304 is the amount by which **locus** is increased for every **IQ** point by which the prisoner exceeds the mean IQ of 88.6.

A further aid to conceptualizing the adjustment process is the representation of the sums of squares of the dependent variable on a Venn diagram. Figure 10.7 shows the Venn diagrams for the unadjusted and adjusted analyses.

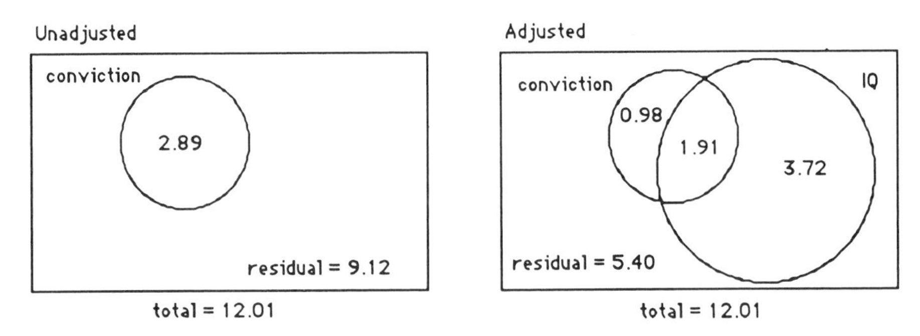

Fig. 10.7 *SS*s for prisoners' locus of control with IQ as covariate.

The sums of squares in the Venn diagram follow directly from the ANOVA summary tables.

In the adjusted Venn diagram in Fig. 10.7 there is an overlap of the regions representing the independent variable and the covariate. This illustrates the portion of the variation in the dependent variable (1.91) explained by **type of conviction** in the unadjusted analysis, but by **IQ** in the adjusted analysis.

In this representation, partialling out **IQ** from both the independent and dependent variables is achieved by erasing the **IQ** region completely, leaving the adjusted *SS*s as follows:

Conviction: 0.98
Residual: 5.40
Total: 6.38

This is exactly the set of *SS*s set out in the adjusted summary in Table 10.15.

Category covariate

The mean **locus of control** scores can be adjusted to what they would be if the same proportion of prisoners in each group had had **sixth form education**. The corresponding adjusted ANOVA summary tables and Venn diagrams can be obtained at the same time. The technique for this follows closely the analysis of covariance discussed for the *continuous* covariate. A statistical computer package is recommended for this. Any package that offers analysis of covariance or general linear modelling should be adequate. See Appendix A.

The summary table showing the adjusted sums of squares is set out in Table 10.17.

Table 10.17 Adjusted analysis of variance (**sixth form** removed from effect of **conviction**)

Source	df_{adj}	SS_{adj}	MS_{adj}	F_{adj}
Conviction	1	3.76	3.76	8.58[a]
Residual	17	7.46	0.4388	
Total	18	11.22		

[a] $F_c = 4.45$ at 0.05
$= 8.40$ at 0.01

Note that the sum of squares for **type of conviction** is increased by adjustment when compared with the unadjusted value in Table 10.14. This is because there is a synergic relationship between **type of conviction** and **sixth form**. In other words, the amount of variation explained by the two variables together is more than the sum of the amounts each explain on their own.

The main consequence of this is that **conviction** changes from significant at the 0.05 level in the unadjusted analysis to significant at the 0.01 level in the adjusted analysis.

The mean **locus of control** scores in the two groups are changed by the adjustment procedure. This is expected since the mean proportion of prisoners having attended **sixth form** differs in the two groups. All mean scores are shown in Table 10.18. The increase in the magnitude of the effect of **type of conviction** on **locus of control** due to adjustment is seen in the mean scores as well as in the sums of squares. The difference between the adjusted means is 0.89 compared to 0.76.

Table 10.18 Means for prisoners' **locus of control**

Group	*Proportion in* **sixth form**	*Mean* **locus**	*Adjusted mean* **locus**
Burglary	40%	4.40	4.4645
Violence	20%	3.64	3.5755

The model fitted by the ANOVA procedure is:

$$\text{Expected score for a randomly sampled subject} = \text{overall mean} + \text{conviction effect} + \text{sixth form effect}$$

$$= 4.02 + \begin{Bmatrix} +0.44 \\ -0.44 \end{Bmatrix} + \begin{Bmatrix} +0.32 \\ -0.32 \end{Bmatrix}$$

(Note: this model ignores the possibility of interactions between **conviction** and **sixth form**. It is sometimes referred to as the **Type II** sums of squares.)

Figure 10.8 shows the Venn diagrams for the unadjusted and adjusted analyses. The sums of squares in the Venn diagram follow directly from the ANOVA summary tables.

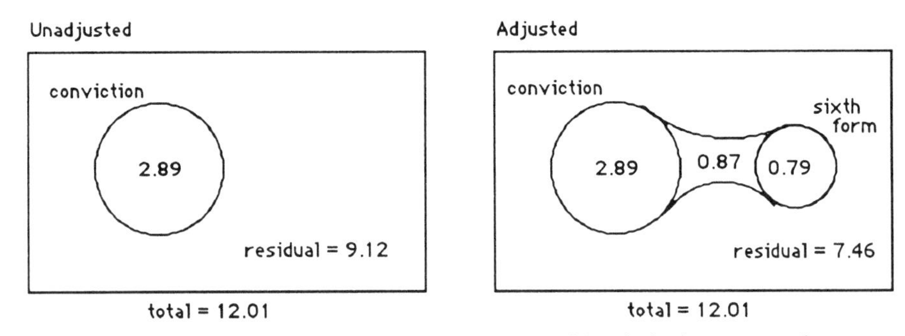

Fig. 10.8 *SS*s for prisoners' locus of control with sixth form attendance as covariate.

In Fig. 10.8, the concave region between the regions representing the independent variable and the covariate represents the synergy (section 10.2.3), that is, the portion of the variation in the dependent variable (0.87) explained only by the combined action of **type of conviction** and **sixth form**.

In this representation, partialling out **sixth form** from both the independent and dependent variables is achieved by erasing the **sixth form** region completely. This leaves the circular region for **conviction** with $SS = 2.89$ and the synergy with $SS = 0.87$. These adjusted *SS*s are set out with the adjusted total as follows:

Conviction: 3.76
Residual: 7.46
Total: 11.22

This is exactly (subject to rounding) the set of *SS*s set out in adjusted form in Table 10.17.

10.3.4 Adjustment for the correlation design

Introduction

Consider the example introduced in section 10.3.2 of a survey in which interest is centred on the effect of **height** on **popularity**. **Sex** and **age** are identified as possible confounding variables because they are each seen as likely to be related to both **height** and **popularity**. Ten males and ten females were included in the study.

The data for the 20 subjects are set out in Table 10.19. **Height** is in metres, **age** in months and the **popularity** scores are higher for the more popular children.

Table 10.19 Data for children's popularity

Height	Popularity	Sex	Age
1.11	8	male	97
1.23	5	male	99
1.18	7	male	102
1.31	14	male	103
1.30	10	male	102
1.25	7	male	100
1.41	6	male	106
1.31	11	male	102
1.15	10	male	101
1.58	13	male	108
1.60	14	female	105
1.31	11	female	97
1.49	13	female	106
1.40	8	female	99
1.63	17	female	107
1.21	13	female	102
1.34	10	female	100
1.20	12	female	100
1.36	9	female	104
1.38	11	female	101

Continuous covariate

The relationship between **height** and **popularity** is measured by the correlation. Its value here is 0.554. It can be interpreted as analysis of variance using the fact that its square is the proportion of the sum of squares of **popularity** explained by **height**. The *SS* of **popularity** is 178.95, of which 30.69% ($0.554^2 = 0.3069$) is explained by **height**. The analysis of variance summary in Table 10.20 displays this result. This is the unadjusted summary table. In Table 10.20, F exceeds F_c at 0.05. This means that the effect of **height** is statistically significant at the 0.05 level.

Age is a continuous covariate. The most comprehensive way to adjust for its effect on the relationship of **height** with **popularity**, since all three variables are continuous, is to use **multiple regression**. This technique is discussed in Chapter 11. However, a convenient technique is available

Table 10.20 Unadjusted analysis of variance

Source	*df*	*SS*	*MS*	*F*
Height	1	54.98	54.98	7.98[a]
Residual	18	123.97	6.887	
Total	19	178.95		

[a] $F_c = 4.41$ at 0.05
$= 8.29$ at 0.01

based on the correlations among the three variables. The technique consists in a formula for partialling out the effect of a continuous variable from a correlation. It is called **partial correlation**. This technique does not make available *SS*s, test of significance or the additive model.

Partial correlation

Let the r symbols stand for correlations of pairs of variables according to the following scheme where variables 1, 2 and 3 refer to **height**, **popularity** and **age**, respectively.

Symbol	*Variables*	*Value*
r_{12}	variable 1 with variable 2	0.554
r_{13}	variable 1 with variable 3	0.745
r_{23}	variable 2 with variable 3	0.487

The formula for the correlation of variable 1 with variable 2, partialling out the effects of variable 3, is

$$\frac{r_{12} - (r_{13})(r_{23})}{[(1 - r_{13}^2)(1 - r_{23}^2)]^{1/2}}$$

This gives

$$\frac{0.554 - (0.745)(0.487)}{[(1 - 0.745^2)(1 - 0.487^2)]^{1/2}} = 0.328$$

for the correlation of **height** with **popularity** while **age** is partialled out. This represents the correlation expected if the children were all of the same **age**. Partialling out **age** reduces the correlation of **height** with **popularity** from 0.554 to 0.328.

The adjusted analysis of variance summary table and the Venn diagram can be obtained, with some manipulation, from the correlations. They are more conveniently obtained from the multiple regression approach (described in Chapter 11). The adjusted analysis of variance summary is in Table 10.21. This shows that the partial correlation is not significant, in contrast to the not-partial correlation.

Table 10.21 Adjusted analysis of variance (**age** removed from effect of **height**)

Source	df_{adj}	SS_{adj}	MS_{adj}	F_{adj}
Height	1	14.73	14.73	2.056[a]
Residual	17	121.78	7.163	
Total	18	136.51		

[a] $F_c = 4.45$ at 0.05
$= 8.40$ at 0.01

The model fitted by the multiple regression procedure is

Expected score for a randomly sampled subject = overall mean + height effect + age effect

$$= 10.45 + 8.91(\text{height} - 1.338) + 0.162(\text{age} - 102.05)$$

The values 1.338 and 102.05 are the means of **height** and **age** respectively. The interpretation is analogous to that of the covariate term in the model discussed in section 7.4. Every unit by which a child exceeds the mean **height** leads to 8.91 extra units of **popularity**, all else being held constant. Every unit by which a child exceeds the mean **age** leads to 0.162 extra units of **popularity**, all else remaining constant.

The Venn diagrams are in Fig. 10.9. As before, the correlation with **age** partialled out is obtained from the adjusted Venn diagram by erasing the whole region of **age**. This leaves sums of squares as follows:

Height: 14.73
Residual: 121.78
Total: 136.51

This shows again the partial correlation as

$$(14.73/136.51)^{1/2} = 0.328$$

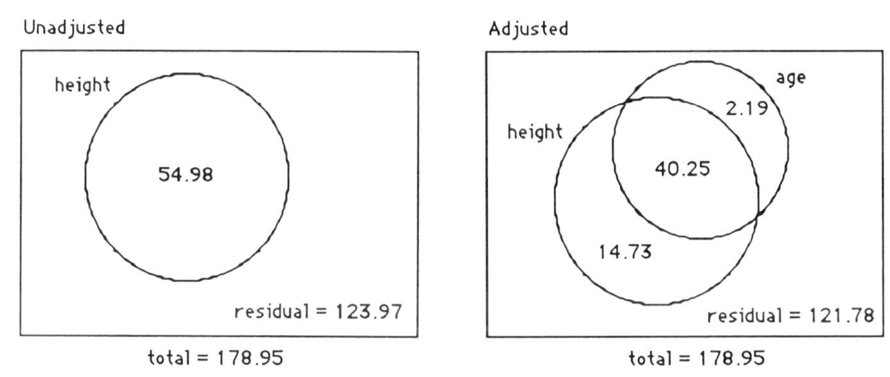

Fig. 10.9 *SS*s for pupils' popularity with age as covariate.

Category covariate

Sex is a category covariate. The most comprehensive way to adjust for its effect on the relationship of **height** with **popularity**, since both the independent and dependent variables are continuous, is to use the **general linear model**. This is an advanced technique. Most books which deal with it require of the reader a considerable background in statistics. An example of the more accessible treatments can be found in Healey (1988).

However, a convenient approach is available based on the analysis of covariance introduced in Chapter 7. The method is applied by supposing the roles of independent variable and covariate are reversed. The standard analysis of covariance computer program then provides the sums of squares needed for drawing the Venn diagram for the fitted model. This can be used just as well for reversed or not-reversed interpretations.

The role-reversed summary table is in Table 10.22. This is the standard application of analysis of covariance. It shows the effect of **sex** on **popularity** with **height** adjusted out.

Table 10.22 Adjusted analysis of variance (role-reversed) (**height** removed from effect of **sex**)

Source	df_{adj}	SS_{adj}	MS_{adj}	F_{adj}
Sex	1	12.25	12.25	1.86[a]
Residual	17	111.72	6.572	
Total	18	123.97		

[a] $F_c = 4.45$ at 0.05
$= 8.40$ at 0.01

It is interesting to note from Table 10.22 that **sex** does not have a significant effect on **popularity** after the effect of **height** has been partialled out. Table 10.22 is used in conjunction with information obtained from the unadjusted analysis of the effect of **sex** on **popularity** (not shown), to draw the Venn diagrams. This is Fig. 10.10.

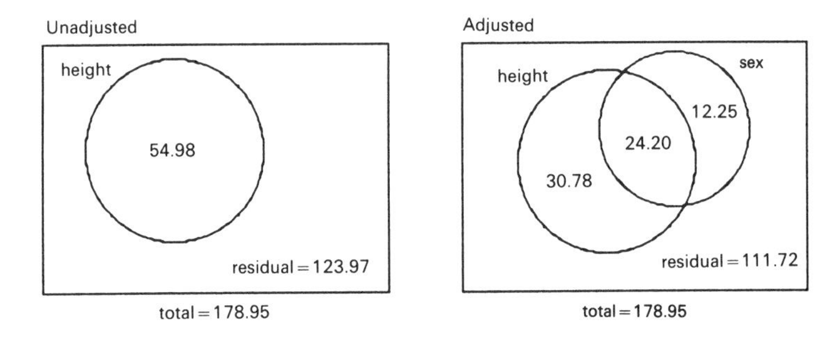

Fig. 10.10 *SS*s for pupils' popularity with sex as covariate.

Table 10.23 Adjusted analysis of variance (**sex** removed from effect of **height**)

Source	df_{adj}	SS_{adj}	MS_{adj}	F_{adj}
Height	1	30.78	30.78	4.684[a]
Residual	17	111.72	6.572	
Total	18	142.50		

[a] $F_c = 4.45$ at 0.05
$= 8.40$ at 0.01

It is a simple task to erase the **sex** region from the Venn diagram to obtain the sums of squares for the effect of **height** on **popularity** adjusted for **sex** as required. The result is in Table 10.23.

From the adjusted sums of squares it is possible to extract the adjusted correlation. It is $(30.78/142.50)^{1/2} = 0.465$. This is the correlation of **height** with **popularity**, with **sex** partialled out. Note that partialling out **sex** has reduced the correlation from 0.554 to 0.465.

The fitted model is:

$$\text{Expected score for a randomly sampled subject} = \text{overall mean} + \text{height effect} + \text{sex effect}$$

$$= 10.45 + 9.267(\text{height} - 1.338) + \begin{Bmatrix} -0.84 \\ +0.84 \end{Bmatrix}$$

The interpretation of the model is as follows: 9.267 is the amount by which **popularity** increases for each unit of **height** in metres by which the pupil exceeds the mean **height** of 1.338 metres. $\mp$ 0.84 is the amount by which **popularity** is reduced below or raised above the mean for being male or female, respectively.

10.4 EXERCISES

10.1 Sixteen small task-oriented groups of junior school pupils were observed in order to explore the effect of 'frequency of production of organizing behaviours' (**org**) on 'quality of completed task' (**quality**).

The proportion of *SS* of the d.v. explained by the i.v. is shown in Fig. 10.11.

Measurements were also obtained on a possible confounding variable, the 'verbal ability scores of the members of each group' (**verbal**). **Verbal** is also related to the d.v. as shown in Fig. 10.12.

The 'synergy' *SS* of 23 units represents the amount by which the *SS* explained by **org** is increased by the presence in the model of **verbal** and vice versa. All the above variables are continuous.

Obtain:

(a) the correlation of **org** with the d.v.
(b) as in (a) but with the effect of **verbal** partialled out.

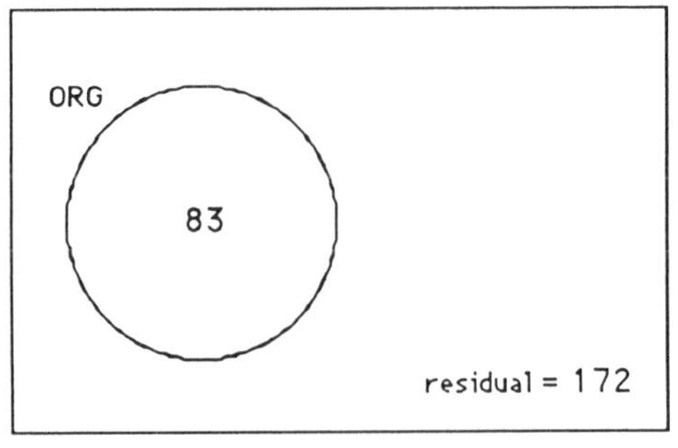

Fig. 10.11

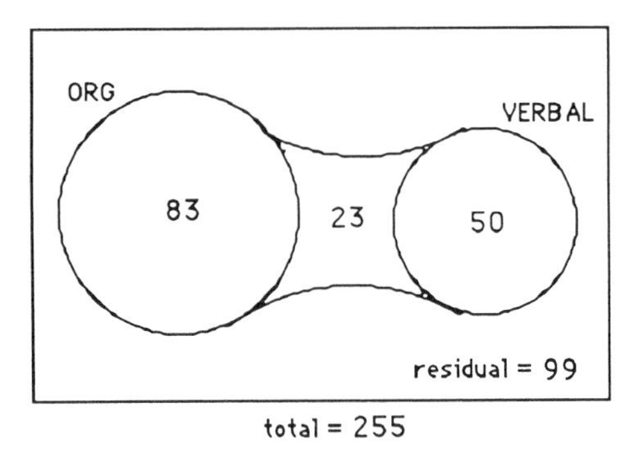

Fig. 10.12

(c) the correlation of **verbal** with the d.v. with the effect of **org** partialled out.

(d) the summary table showing the effect of **org** on the d.v. both adjusted and not adjusted for **verbal**.

Measurements were also obtained on another possible confounding variable, the 'non-verbal ability scores of the members of each group' (**nonverbal**). **Nonverbal** is a continuous variable. The relationship of **nonverbal** to the d.v. is shown in Fig. 10.13.

Obtain:

(e) the correlation of **org** with the d.v. with the effect of **nonverbal** partialled out.

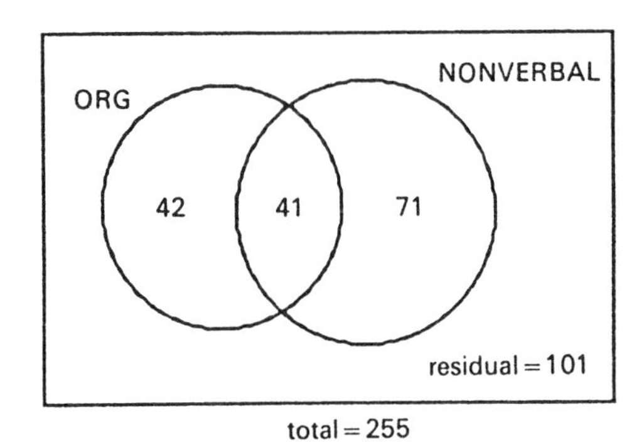

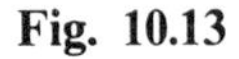
Fig. 10.13

(f) The summary table showing the effect of **org** on the d.v. both unadjusted and adjusted for **nonverbal**.

10.2 Use unweighted means analysis to obtain the means adjusted for the category type confounding factor in the following non-randomized experiment.

Thirteen male and 15 female pupils were coached in basket-ball for one hour according to either the traditional method or an imaging method. Successful shots (**baskets**) out of 20 attempts are set out in the table below. **Sex** is regarded as a confounding factor. The individual pupils could not be randomized.

	Male				Female		
Traditional	15	17	12	13	10	11	6
	15	8	13	19	8	5	7
Imaging	10	9	10		6	9	4
	8	7			10	5	6
					7	4	6

(a) Compare the size of the effect of the conditions in the adjusted and unadjusted analyses.
(b) If the resulting two-factor ANOVA resulted in Fig. 10.14, set out the adjusted and unadjusted summary tables, complete the corresponding tests of significance and discuss the effect of the adjustment.

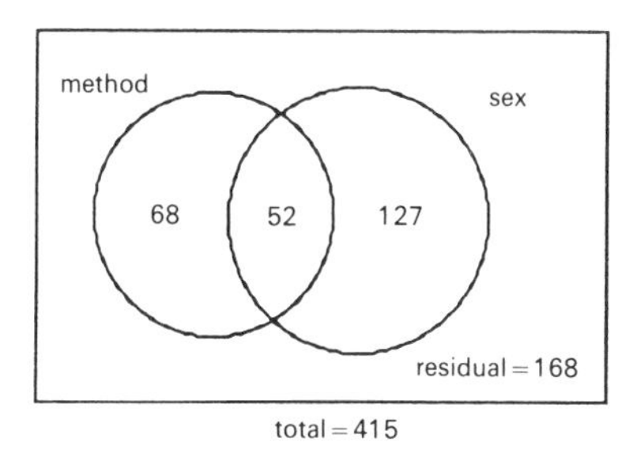

Fig. 10.14

10.3 Set out below are the number of successful basketball shots (**baskets**) out of 100 attempts by pupils trained by either method **A** (traditional) or by method **B** (imaging). The covariate is the number of **months** the pupil had been playing basketball.

Raw data:

Group **A**	**baskets** (Y)	15	42	25	12	3
	months (X)	23	35	29	20	17
Group **B**	**baskets** (Y)	17	52	28	26	38
	months (X)	10	25	15	12	20

Figure 10.15 displays each pupil's result as a point labelled ⊡ for method **A** and ◆ for method **B**.

(a) Calculate and plot each group mean on Fig. 10.15.
   Roughly draw the best fitting parallel regression lines passing through the group means.
   Use these lines to estimate what each group mean would be if all pupils had been playing for 20 months.
(b) Display the unadjusted and adjusted effects of the conditions as bar charts. Has the adjustment increased or reduced the size of the effect?
(c) Use Fig. 10.16 to construct an adjusted summary table and to carry out an F-test of the adjusted effect of the conditions.

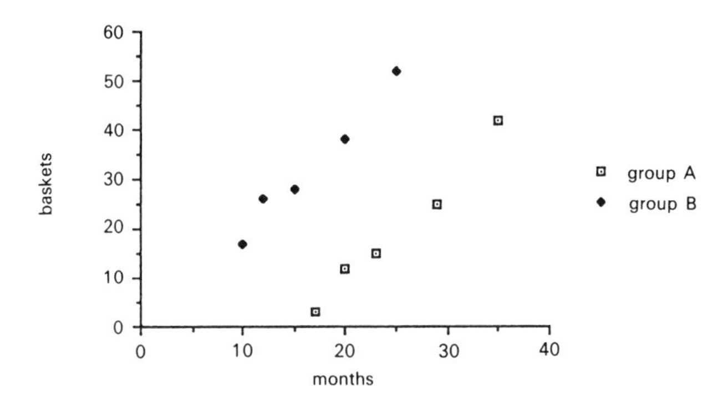

Fig. 10.15 Baskets versus months for two groups.

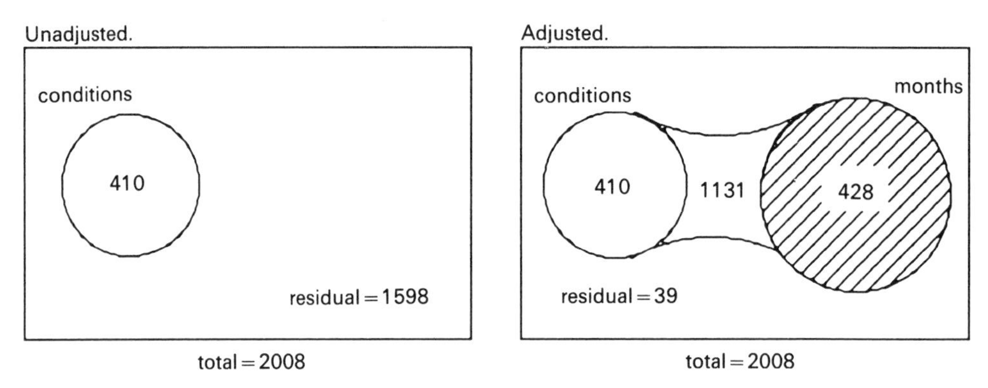

Fig. 10.16 Venn diagram for SSs.

10.4 The following four empirical investigations are to be carried out by use of a survey or by comparison of pre-existing groups. Consider (a)–(d) below for each investigation.

(a) Identify the indpendent and dependent variables and note for each whether it is category or continuous type and for the i.v. whether it is intrinsic or extrinsic.

(b) Identify possible confounding variables and identify them as category or continuous type.
(c) Suggest ways in which the effects of the confounding variables might be taken account of.
(d) Consider the feasibility of using a randomized experiment.

Investigation 1
You are to research the effect of 'height of child' in a local junior school on peer-rated 'popularity scores'. The measure of popularity will be on a scale from 0 to 20. Correlation will be used to analyse the data.

Investigation 2
It is required to investigate the effectiveness of a recent TV programme about vaccination. Mothers registered at a particular clinic who were due to have their child vaccinated around the date of the broadcast are to be interviewed.

Whether or not mothers saw the programme will be related to whether or not they had the vaccinations done in the three weeks following the broadcast.

Investigation 3
'Pupil-centred' and 'traditional' teaching methods are to be compared. Two equivalent classes in Hatfield Green School were selected and randomly assigned to one of each of the methods for all English lessons for a year. At the end of the year a t-test was used to compare the mean scores on an English test.

Investigation 4
The Home Office has commissioned you to research the effect of type of conviction on 'locus of control' scores on prisoners. The two types of conviction to be considered are 'violence' and 'burglary'. Locus of control is to be measured using a 7-point scale on prisoners who have reached the end of their first four weeks of imprisonment.

11 Multiple regression

11.1 INTRODUCTION

Other sections of this book are concerned with designs in which at least one of the independent variables is of category type. Multiple regression refers to designs in which all independent variables (i.v.s) are continuous. Its inclusion is justified by its similarity to the two-factor unbalanced design and the analysis of covariance; this contributes to completeness without requiring many new concepts.

Table 11.1 summarizes the structures of the designs based on two i.v.s dealt with here and in Chapters 6 and 7.

Note that in this chapter interaction between i.v.s is omitted from all analyses even if it is present in the data.

Table 11.1 The structures of various models

Model	*Dependent variable type*	*Independent variable type*
Two-factor ANOVA	Continuous	Both category
One-factor ANCOVA[a]	Continuous	1 category 1 continuous
Two-variable multiple regression	Continuous	Both continuous

[a]Regarding the covariate as an i.v.

11.2 OVERVIEW OF DESIGNS, VARIABLES AND ORTHOGONALITY

It is unusual that a continuous variable is varied experimentally. An exception is an experiment in which individuals are exposed to a tone whose pitch is set according to a random process. The dependent variable is the difference between the pitch to which the subject was exposed and the pitch of the note produced by the subject in attempting to reproduce it.

More commonly a continuous i.v. arises as a measurement in a survey or observational study where it is more often intrinsic rather than extrinsic. In this form the contitinuous i.v. has already been presented as a continuous covariate in Chapter 7.

The interconnections between types of i.v. and properties and names of designs is represented on the Euler diagram in Fig. 11.1. The labelling of the regions (a)–(e) refers to the explanations that follow. Note that a factorial design consists of category i.v.s and a continuous d.v.

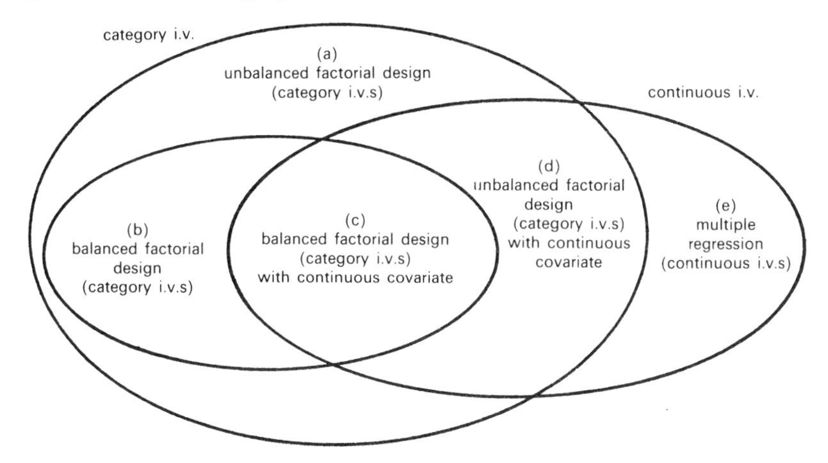

Fig. 11.1 Arrangements of types of variables and designs.

(a) *Unbalanced factorial design* This arises either from a randomized experiment with unequal numbers in the cells or from a survey. The *SS*s of this design are represented in the Venn diagram in Fig. 11.2(a). This was introduced in section 10.2.

(b) *Balanced factorial design* This usually arises from a designed experiment with or without randomization. It has identical numbers in the cells and leads to an orthogonal analysis of variance. This means its *SS*s can be represented by *non-overlapping* regions of the Venn diagram (Fig. 11.2(b)). This was discussed in Chapter 6.

(c) *Balanced factorial design with covariate* This is the same as (b), but the addition of a continuous covariate introduces **non-orthogonality**. The *SS* for the covariate is expected to overlap the other regions of the Venn diagram (Fig. 11.2(c)). This was introduced in Chapter 7 for the single i.v. case.

(d) *Unbalanced factorial design with covariate* This is the same as the design in (a), but with a continuous covariate whose *SS* on the Venn diagram representation is expected to overlap all other regions (Fig. 11.2(d)).

(e) *Multiple regression* Continuous covariates have mutually overlapping regions on the Venn diagram. For simplicity it is supposed here that there are no interactions among the covariates. This design usually arises from a survey (Fig. 11.2(e)).

11.3 COMPARISON OF MODELS WITH CATEGORY AND CONTINUOUS INDEPENDENT VARIABLES

11.3.1 Overview

In this section similarities are illustrated between models for category and continuous type i.v.s. The similarity of these models is implicit in their

(a) Unbalanced factorial design.

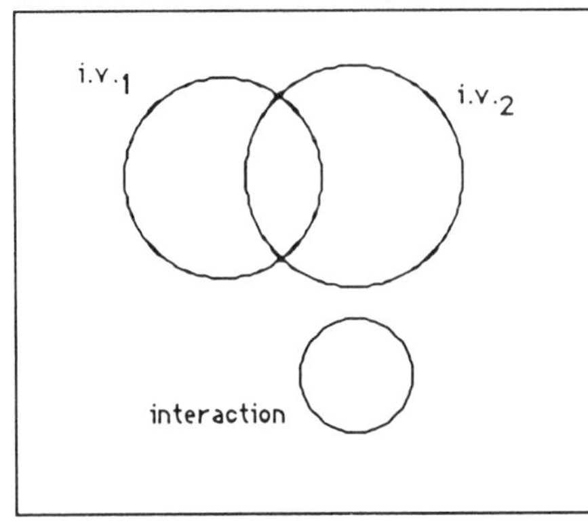

(b) Balanced factorial design.

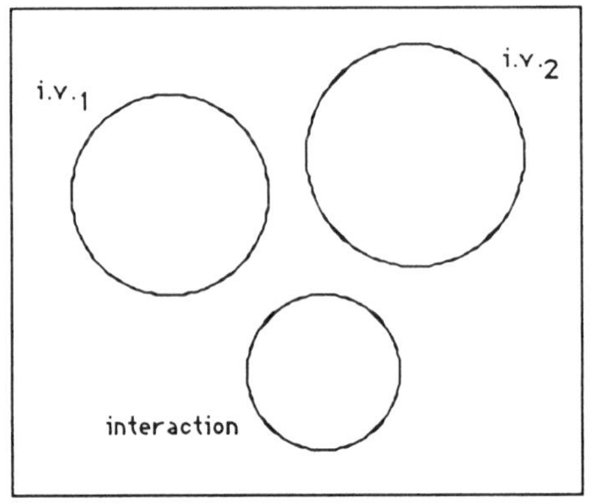

(c) Balanced factorial design with covariate.

(d) Unbalanced factorial design with covariate.

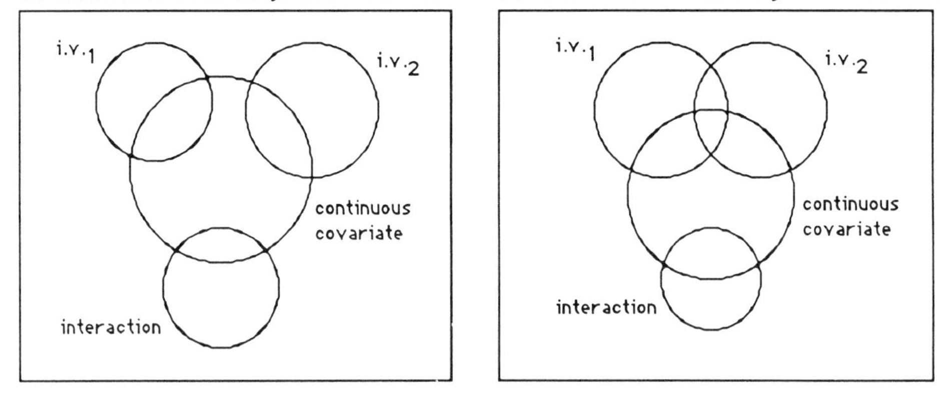

(e) Multiple regression.

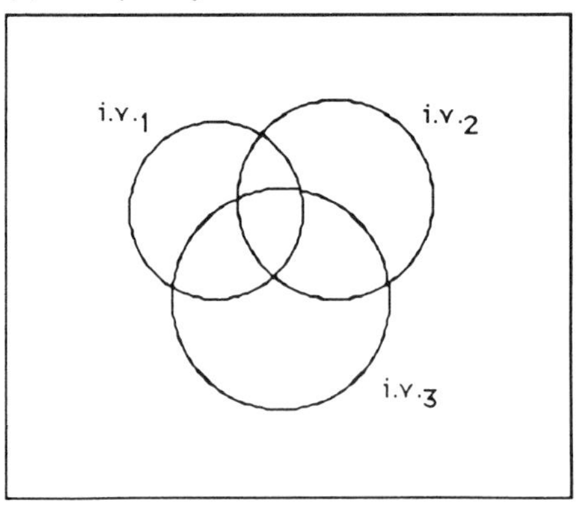

Fig. 11.2 Venn diagram representation of sums of squares of various models.

being members of the class of models called **general linear models.** Multiple regression is introduced by showing it as a version of the models already developed earlier in the text. In particular, the reader is referred to section 10.2 (two-factor unbalanced design) and the section on adjustment for a continuous covariate in section 10.3.4.

Refer to Table 11.2.

(a) *Summary* The designs are briefly summarized. Note that the balanced two-factor design is the same example as that used in Chapter 6.

(b) *Raw data* The last column represents the d.v. in both tables of data.

(c) *Models* Note that no interaction is present in the models. By choosing to ignore any interaction between the variables a greater portion of the *SS*s remains as unexplained **residual**.

The terms in the balanced two-factor model represent the deviations due to particular levels of the factors. For example '+4' is the amount by which **sewing times** are increased when **machine type 3** is used.

The terms in the multiple regression model represent the deviations due to particular values of the independent variables. For example 0.73 (**professional motivation** −11.0) is the amount by which **exam grade** is increased for an individual with a particular value of **professional motivation**. The gradient or scale ratio of 0.73 is the amount by which **exam grade** is increased for a unit increase in **professional motivation** supposing other i.v.s are held constant.

The terms 0.73 and 0.639 are known as **regression coefficients** or **multiple regression coefficients**.

The multiple regression model is often presented in a modified but equivalent form in the output from computer programs. The alternative presentation is obtained by simple algebraic manipulation. The equation for exam grades is rewritten:

$$\text{expected exam grade} = [10.0-(0.73)(11.0)-(0.639)(11.33)] + {} + 0.73\ (\text{professional motivation}) + 0.639\ (\text{coursework})$$

which reduces to:

$$\text{expected exam grade} = -5.27 + 0.73\ (\text{professional motivation}) + 0.639\ (\text{coursework})$$

The reverse transformation can easily be carried out provided the mean values of the i.v.s are known.

(d) *ANOVA* For the multiple regression design the two sequential orderings of independent variables lead to different partitions of the total variation. In the table with **professional motivation** in the first position, the *SS* and test of significance refer to the *entire* contribution of **professional motivation** to explaining variation in **exam grades**. The *SS* of 35.0 corresponds to the entire circular region of the Venn diagram. **Professional motivation** has had the 'first bite at the cake'.

In the table with **professional motivation** in second position, the *SS* of 10.95 corresponds to the half-moon shaped region of the Venn diagram. **Professional motivation** has only been able to explain variation in **exam grade** left unexplained by **coursework**. In this table the *SS* and test of significance refer to the unique contribution of **professional motivation**.

The continuous i.v.s in the multiple regression have one degree of freedom associated with each of them.

Table 11.2(a) Experiment with two category-type i.v.s

(a) *Summary of design* Sixty machinists were randomly allocated, five to each combination of **type of machine** and **training**. Performance was measured as **completion time** of a sewing task.

(b) *Raw data*

Subject no.	**Machine**	**Training**	**Completion time**
1	1	1	12.5
2	1	1	11.5
3	1	1	13.9
4	1	1	12.8
5	1	1	9.3
6	1	2	13.1
7	1	2	12.5
⋮	⋮	⋮	⋮
60	3	4	13.4

(c) *Model*

$$\text{Expected completion time} = 15 + \underbrace{\begin{Bmatrix} 0 \\ -4 \\ +4 \end{Bmatrix}}_{\textbf{machine}} + \underbrace{\begin{Bmatrix} 0.33 \\ -2.33 \\ 0.33 \\ 1.67 \end{Bmatrix}}_{\textbf{training}}$$

(d) *ANOVA summary table*

Source	*SS*	*df*	*MS*	*F*
Machine	640.0	2	320.0	27.8
Training	126.7	3	42.2	3.66
Residual	622.7	54	11.5	
Total	1389.4	59		

(e) *Sizes of effects* **Machine** explains 640 out of 1389 = 46% of total variation in completion times.

Training explains 126.7 out of 1389 = 9% of total variation in completion times.

(f) *Venn diagram*

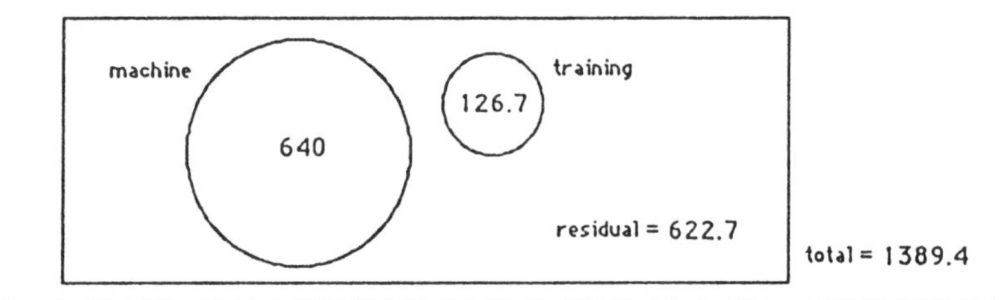

(e) *Sizes of effects* These need no further explanation in the two-factor category i.v. design. In the multiple regression design the size of effect can refer to either the whole or the unique effect of the variable.

Thus 10.95/101.94 = 0.107 is the proportion of variation in **exam grade**

Table 11.2(b) Survey with two continuous-type i.v.s (multiple regression)

(a) *Summary of design* A survey was carried out on six students to relate their final **exam grades** to an attitude measure (**professional motivation**) and to a measure of performance during the course (**mean coursework mark**).

(b) *Raw data*

Subject no.	**Professional motivation**	**Course work**	**Exam grade**
1	14	19	18
2	11	8	9
3	8	14	8
4	13	10	8
5	10	8	5
6	10	9	12

(c) *Model*

expected exam grade = 10.0 + 0.73 (**professional motivation** − 11.0) + 0.639 (**coursework** − 11.33)

(d) *ANOVA summary tables* (sequential *SS*s i.e. Type I)

Source	*SS*	*df*	*MS*	*F*
Professional motivation	34.98	1	34.98	3.123
Coursework	33.36	1	33.36	2.98
Residual	33.60	3	11.2	
Total	101.94	5		

Source	*SS*	*df*	*MS*	*F*
Coursework	57.39	1	57.39	5.12
Professional motivation	10.95	1	10.95	0.98
Residual	33.60	3	11.2	
Total	101.94	5		

(e) *Sizes of effects* **Professional motivation** explains 34.98 or 10.95 out of a total variation of 101.94. This represents 34.3% or 10.7% depending on sequential position.

Coursework explains 33.36 or 57.39 out of a total variation of 101.94 = 32.7% or 56.3% depending on sequential position.

(f) *Venn diagram*

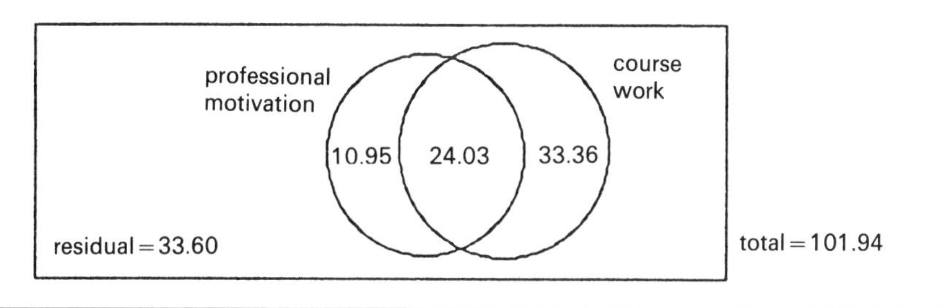

explained by **professional motivation** with **coursework** partialled out. 0.107 is the square of the **semi-partial correlation**. This gives $(0.107)^{1/2} = 0.327$ as the correlation of **professional motivation** with **exam grade** with **coursework** only partialled out of **professional motivation** (hence semi-partial correlation).

Note that

$$\frac{10.95}{10.95+33.60}=0.246$$

is the square of the **partial correlation**. The square root of this, 0.496, is the correlation of **professional motivation** with **exam grade** with **coursework** partialled out.

(f) *The Venn diagram* The *SS* of the i.v.s **machine** and **timing** in the balanced design are represented by non-overlapping regions. By contrast, the *SS*s of the i.v.s **professional motivation** and **coursework** in the multiple regression overlap.

The overlap represents **interdependence** of the i.v.s.

11.3.2 Comparison of the models

Table 11.3 represents the components of the models for each design as graphs. In each model a fictitious individual's score (represented by a single point) is used for illustration purposes.

The two-factor balanced design is represented in Table 11.3(a). The individual selected is in the group that experienced **machine type 3** and obtained a score above the mean for the group (i.e. was slower at sewing). According to the model the score is made up of the following components (set out here in the same order as in the graph):

error/residual (due to individual characteristics,
measurement error, the other factor or interaction)

+

deviation from overall mean completion time due to
machine 3 (+4)

+

overall mean **completion time** (15)

In the multiple regression model the individual selected has an above average value for **professional motivation** and obtained a grade on the **exam** above that expected from his or her **professional motivation** value. According to the model the score is made up of the following components (set out here in the same order as in the graph):

error/residual (due to individual
characteristics or measurement
error or the other i.v.)

+

individual's deviation from overall
mean **exam grade** due to **professional motivation** 0.73(**professional motivation** − 11)

+

overall mean **exam grade** (10)

For both models a second graph would be required to represent the effect of the second independent variable.

Table 11.3 Components of models (**p.m. professional motivation; c.w. coursework**)

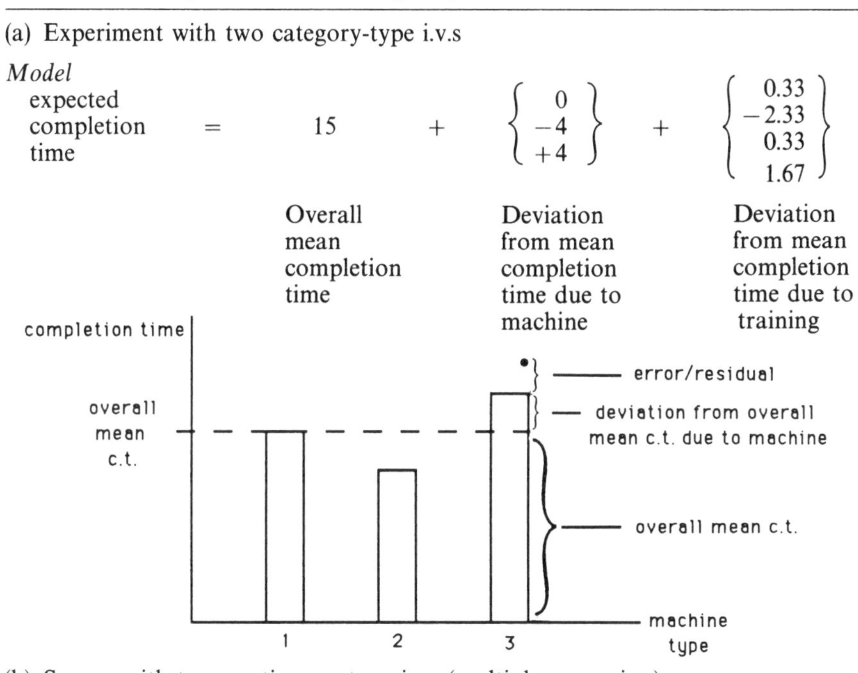

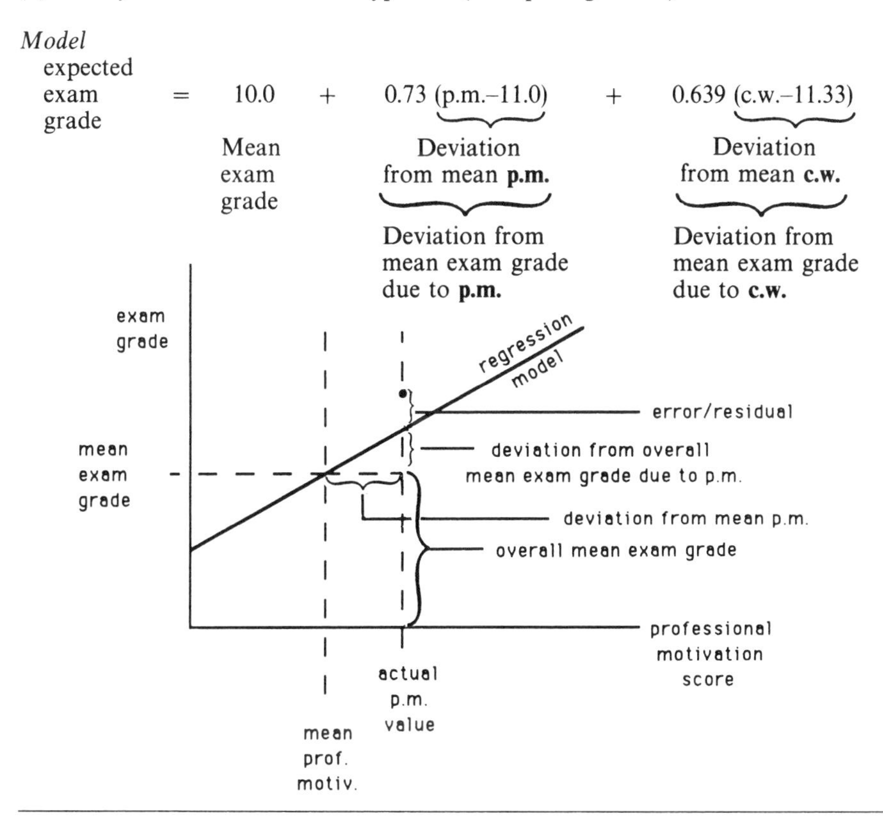

11.4 GLOSSARY OF TERMS FOR MULTPLE REGRESSION

Goodness of fit	Proportion of variation explained by regression model (same as *R*-squared).
Independent variables	Also known as predictors or as explanatory variables (all continuous-type).
Partial correlation	Partial correlation of an i.v. with the d.v. is the correlation with both adjusted for all other i.v.s. In other words, it is the correlation between an i.v. and the d.v. in a world in which all subjects have identical values for all other i.v.s.
Regression coefficient	Weight or scale multiplier for an independent variable chosen so as to maximize the proportion of variance explained by the regression model.
Regression equation	Formula enabling calculation of predicted value of the dependent variable from values of the independent variables.
Relationship of scale	Interpretation of regression coefficient as scale of relationship between the dependent variable and the corresponding independent variable when all other independent variables are held constant.
Residual	Difference between an observed value and the corresponding predicted value of the dependent variable.
Residual variance	Variance across the sample in values of the dependent variable not explained by the regression model.
R-squared	Proportion of variation in the dependent variable explained by all the independent variables in the model.
Semi-partial correlation	Semi-partial correlation of an i.v. with the d.v. is the correlation with the i.v. only adjusted for all other i.v.s. It is a measure of the relationship between the part of an i.v. not explainable by the other i.v.s and the d.v. In other words, it is a measure of the unique contribution of an i.v. to explaining the variation in the d.v.
Sequential inclusion	First i.v. on list has first bite at cake, second only has possibility to explain variation not already explained by the first etc. (hence **sequential** *SS*).
Stepwise inclusion	Automatic algorithm: first i.v. to be included is the best single predictor, second is best at explaining variation not already explained by the first, etc. (a stopping rule usually applies).
Unique contribution	Portion of the variation in the dependent variable explained by an independent variable and not explainable by any other i.v.s.

11.5 SEQUENTIAL MODEL CONSTRUCTION

11.5.1 Introduction

In survey research there is often a need to carry out exploratory analysis. The question which concerns the researcher is, 'Which combination of i.v.s best explains variation in the d.v.?'.

Implicit in this question is a requirement for **parsimony**. This means explaining as much variation as possible using as few i.v.s as possible. One approach to this involves fitting all possible models in a **hierarchical sequence**.

Step 1 Fit all models involving a single i.v.
Step 2 Fit all models involving pairs of i.v.s
Step 3 Fit all models involving three i.v.s

and so on. If there are no more than four i.v.s this approach is manageable. It involves fitting $3+3+1=7$ models for the three-variable model and $4+6+4+1=15$ models for the four-variable model, and so on. The number of models is obtained by counting all combinations of i.v.s in the model.

11.5.2 Selecting the best model

The criterion for model selection is the proportion of variation in the d.v. explained. This is referred to as **multiple correlation squared** or **R-squared**.

Consider as an example a multiple regression analysis involving three independent variables:

X_1 average **excess pulse rate** above resting level
X_2 intake of **caffeine**
X_3 **body weight**

The aim of the study is to explore the extent to which these three i.v.s explain variation in individuals' **reaction times**. **Reaction time** is the dependent variable. Data were obtained from a sample of 30 individuals.

The proportions (as percentages) of total *SS* explained by each model are set out in Table 11.4. The best models involving one and two i.v.s are the ones which explain 35.1% and 52.3% respectively.

Table 11.4 Proportions of variation in reaction times explained by combinations of three independent variables

Independent variables included in regression	*Multiple* R^2 (%)
X_1	18.2
X_2	26.2
X_3	35.1
X_1X_2	37.3
X_1X_3	46.6
X_2X_3	52.3
$X_1X_2X_3$	60.6

11.5.3 Test of significance

The further question is, 'Is the best model one involving one, two or three independent variables?'.

The decision depends on circumstances. Clearly the **full model**, that is the model involving all i.v. s, explains the most variation in the d.v. However, it may be that the additional amount of variance that it explains, beyond that explained by the best of the models involving one fewer i.v. s, is not significantly different from zero. The hypothesis test formulation is as follows:

H_0: no additional variance explained
H_1: some additional variance explained

It can be seen from the Venn diagram in Fig. 11.3 that the required test is of the unique contribution of variable X_1 to explaining variation in the d.v. This is because X_2 and X_3 form the best pair. The unique contribution of X_1 is an SS of 26. The summary table for the test is in Table 11.5.

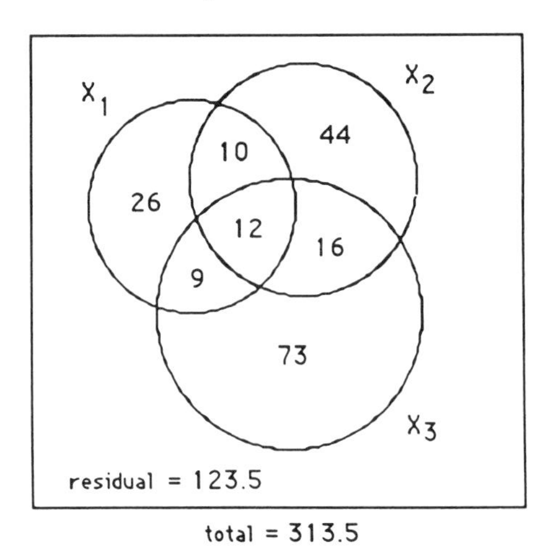

Fig. 11.3 Sums of squares for the reaction time example.

Table 11.5 Summary table for test of unique contribution of variable X_1

Source	*df*	*SS*	*MS*	*F*
X_2 and X_3	2	164	82.0	17.26
X_1 (unique)	1	26	26.0	5.47[a]
Residual	26	123.5	4.75	
Total	29	313.5		

[a] $F_c = 4.22$ at 0.05

The table is constructed according to the principles introduced in Chapter 10. The degrees of freedom are one for each continuous i.v. and one less than the sample size for the total SS. The degrees of freedom for the residual SS are equal to the difference between the total df and the df for terms fitted in the model.

The SSs in Table 11.5 are taken directly from Fig. 11.3. The SS (unique) for variable X_1 is the difference between the SS explained by the full model (all i.v.s involved) and the SS explained by the model based on variables X_2 and X_3.

In this example the unique contribution of X_1 to explaining variation in reaction times is statistically significant at the 0.05 level. Therefore the model involving all three i.v.s is, in this sense, a better model than the one which only involves X_2 and X_3.

11.5.4 Automatic stepwise procedures

Most computer packages that provide multiple regression offer **forward stepwise procedures**. These are procedures that at the first step identify the best single **explanatory variable** and bring in, step by step, further variables, choosing one at a time according to the amount of variation (SS) additionally explained. The result is usually expressed in terms of increments in multiple R-squared. The process is here set out in detail.

Step 1 Identify the best single explanatory variable, carry out the test of significance, report the value of R-squared and the coefficients for the model.

Step 2 Identify the i.v., not previously involved, that explains the greatest amount of the variation left unexplained after the previous step. Carry out the test of significance of the newly explained variation, report the value of R-squared and the coefficients for the model.

Steps 3, 4 etc. Further steps are repeats of step 2.

The model that is fitted sequentially by this procedure for the reaction time study is represented by the Venn diagram in Fig. 11.4. In the reaction time example the i.v.s are included in the order X_3, then X_2, then X_1.

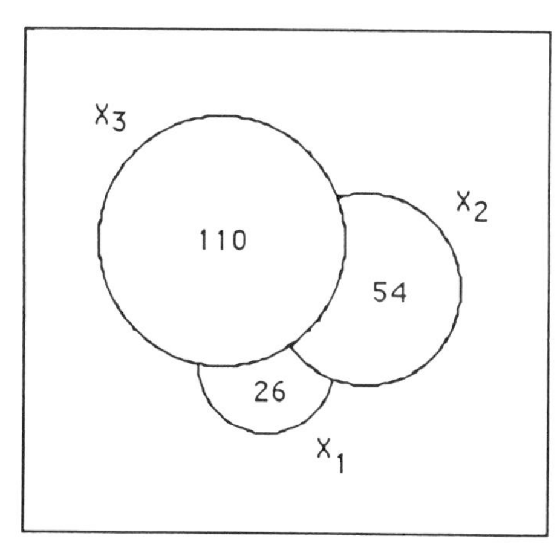

Fig. 11.4 Sequential stepwise fitting of regression model.

Stopping rule The automatic procedure for building up a sequential model usually includes an automatic stopping rule. A common rule is to stop adding terms to the model as soon as the last statistically significant term has been added.

11.6 EXERCISES

11.1 A survey was carried out on five management trainees to investigate the relationship between the scores obtained in a personality test, an aptitude test and the final assessment.

Subject	Personality	Aptitude	Final assessment
1	48	72	69
2	46	68	58
3	60	80	81
4	42	71	58
5	50	82	84
Mean	49.2	74.6	70.0

Figure 11.5 shows the Venn diagram for *SS* for **final assessment**. Multiple regression coefficients of 0.35 for **personality** and 1.68 for **aptitude** were obtained from a computer analysis.

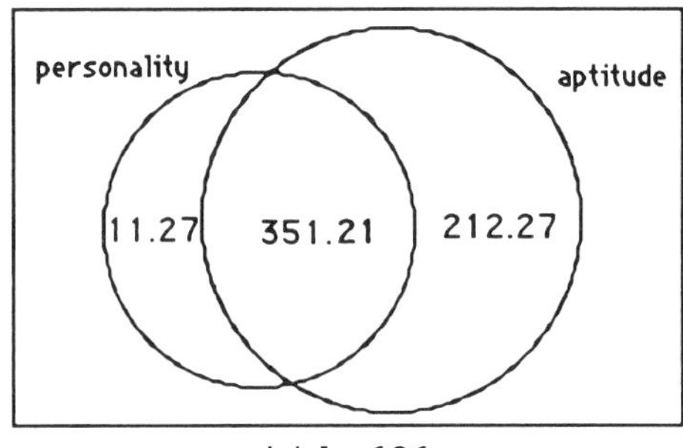

Fig. 11.5

(a) Use the information given above to write out the model for predicting a subject's **final assessment**.

(b) Use the model to estimate the **final assessment** for each of the five **subjects**. Calculate the residuals and hence the sum of squares of residuals. Check your answer by reference to the Venn diagram.

(c) Construct an ANOVA table for a sequential analysis taking **personality** as the first effect. Complete the F-tests and report your conclusions.

(d) Obtain multiple R-squared and explain what it means in the context of this investigation.

Analysis for Further Experiment Designs

Two-factor designs with between- and within-subjects factors

12.1 INTRODUCTION

The researcher should decide on within-subjects or between-subjects for each factor in an experiment on the basis of the discussion in section 9.5 and common sense. Previous experiments in the same general area should not be copied thoughtlessly.

While the general principles of the two-factor design were established in Chapter 6 for the two-factor independent groups design (i.e. two between-subjects factors or between–between or BB), some important differences of detail need to be considered when using the between–within (BW) and within–within (WW) designs.

Examples of BW and WW designs are introduced in sections 12.2 and 12.3. Attention needs to be given to the construction of the specific tests of hypotheses and calculation of size of effects.

12.2 EXAMPLE OF A BW DESIGN

12.2.1 Introduction

Individuals were randomly allocated to three groups of 10. Each group experienced a different condition (**rounded**, **straight** or **irregular**) of visual stimulus.

Each stimulus was presented in both a **control** and an **experimental** version to each subject. This is a version of the well-known psychology experiment on the Stroop effect.

The factors were:

stimulus (3 levels, between-subjects)
condition (2 levels, within-subjects)

The dependent variable was reaction time in milliseconds. It was obtained twice for each subject, once in the control condition and once in the experimental condition.

The raw data and mean reaction times are in Tables 12.1 and 12.2.

Table 12.1 Reaction times data for BW design (milliseconds)

Rounded			**Straight**			**Irregular**		
Subject	**Control**	**Expt.**	*Subject*	**Control**	**Expt.**	*Subject*	**Control**	**Expt.**
1	9.00	7.00	11	8.00	11.00	21	9.36	13.70
2	10.00	12.00	12	7.00	10.00	22	16.09	13.08
3	10.41	12.38	13	11.80	16.52	23	7.17	9.94
4	9.66	8.06	14	12.56	12.62	24	7.81	8.34
5	6.14	7.00	15	26.96	36.62	25	9.31	13.78
6	11.81	8.51	16	7.83	7.19	26	10.82	9.56
7	8.31	8.42	17	5.69	9.46	27	7.19	7.77
8	16.01	8.92	18	12.99	13.49	28	7.31	6.66
9	6.74	8.18	19	6.62	10.02	29	8.80	10.20
10	10.25	6.61	20	7.78	6.50	30	5.24	7.53

Table 12.2. Mean reaction times for BW design

		Condition		
		Control	**Experimental**	Mean
Stimulus	**Rounded**	9.83	8.71	9.27
	Straight	10.73	13.34	12.04
	Irregular	8.91	10.06	9.48
	Mean	9.83	10.70	10.26

The model follows from the table of means. It is:

$$\text{Expected value of reaction time} = \underbrace{10.26}_{\text{Overall mean}} + \underbrace{\left\{\begin{matrix} -0.43 \\ +0.43 \end{matrix}\right\}}_{\text{Effect of } \textbf{condition}} + \underbrace{\left\{\begin{matrix} -0.99 \\ +1.78 \\ -0.78 \end{matrix}\right\}}_{\text{Effect of } \textbf{stimulus}} + \underbrace{\left\{\begin{matrix} +0.99 & -0.99 \\ -0.88 & +0.88 \\ -0.14 & +0.14 \end{matrix}\right\}}_{\text{Effect of interaction}}$$

The main use of the model is for comparing the sizes of the effects of the different terms in the model in units of the dependent variable. Here it shows that the effect of **condition** is ± 0.43 milliseconds whereas the effect of **stimulus** is of the order -1.0 to $+1.8$ milliseconds.

However, since the numerical values of the model usually have to be calculated by hand, they are not often obtained.

No new issues arise in presenting the effects of the independent variables and interaction in terms of the cell and marginal means. They are presented as bar charts and interaction diagram in Fig. 12.1.

It is evident that in this example reaction times are longest for the straight level of **stimulus** and for the experimental level of **condition**. There is evidence of an interaction shown by the non-parallel relationship between the lines linking cell means in the same condition. However, an analysis of variance is required before statements can be made about the presence of effects in the population.

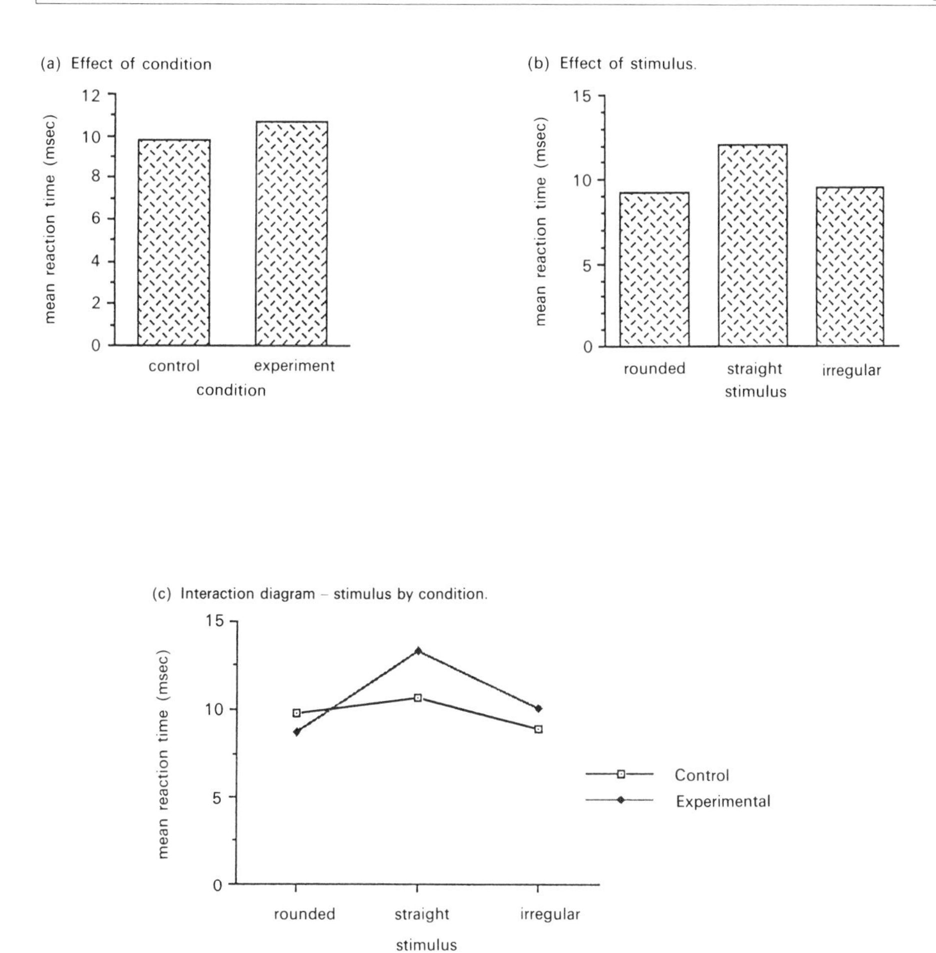

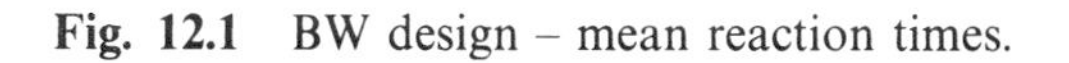

Fig. 12.1 BW design – mean reaction times.

12.2.2 Analysis of variance for BW design

The summary table for analysis of variance is given in Table 12.3. The rules for the analysis of variance of the BW design will be dealt with in section 12.4. However, there are two points worth making here. The first point

Table 12.3 Analysis of variance for the BW design

Source	*df*	*Type*	*SS*	*MS*	MS_{error}	F	F_c
Stimulus	2	B	94.8	47.4	43.2	1.10	3.35
Condition	1	W	11.5	11.5	4.2	2.74	4.21
Stimulus × Condition	2	W	35.4	17.7	4.2	4.21	3.35
Subjects (within-groups)	27		1166.4	43.2			
Subjects × Condition	27		113.4	4.2			
Total	59						

F_c is the 0.05 critical value

is the list of sources of variation appearing in the table. It is helpful to remember that this design is a combination of the one-factor between-subjects design of Chapter 4 and the one-factor within-subjects design of Chapter 5. If **subjects** is relabelled **within groups** and **subjects × condition** is relabelled **reliability** it is seen that **stimulus** together with its error term **subjects** represent the between-subjects dimension, whereas **condition** with its error term **subjects × condition** represent the within-subjects dimension.

The second point is the ten-fold size difference (43.2 compared to 4.2) between the MS_{error} term of **stimulus**, the between-subjects factor, and the MS_{error} term of **condition**, the within-subjects factor. It is evident in this example that an effect has to be much larger to attain significance if it is between- rather than within-subjects. This reinforces the discussion in section 9.5.2.

12.2.3 Tests of significance for the BW design

The basic analysis of variance in Table 12.3 is obtainable from most of the well-known statistical packages for computers (Appendix A). It makes available tests of hypothesis about the existence in the population of effects due to the two factors and their interaction. It is seen that only in the case of the interaction does the F value exceed the 0.05 critical value. Thus H_0 is rejected for interaction but not rejected for **stimulus** and **condition**.

12.2.4 Venn diagram for the BW design

The existence of two orthogonal dimensions to the analysis of variance justifies the use of a two-part Venn diagram to represent it diagrammatically (Fig. 12.2).

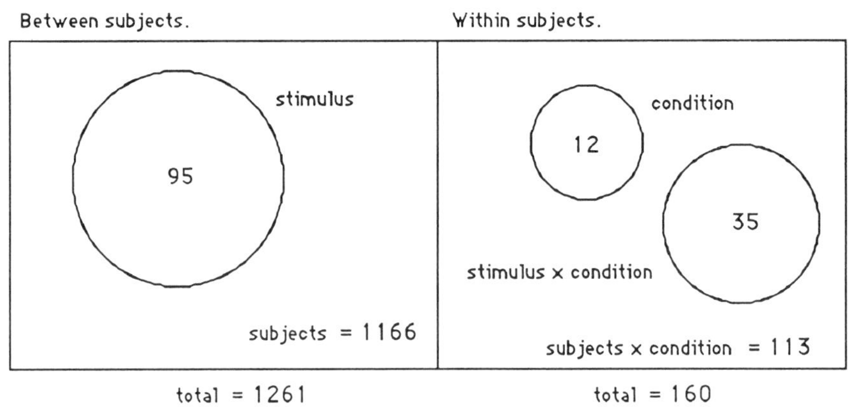

Fig. 12.2 Venn diagram for analysis of variance for BW design.

12.2.5 Size of effect for the BW design

Because of the existence of distinct 'between' and 'within' dimensions in the BW design, the size of an effect is expressed as a proportion of the appropriate total SS.

Therefore the proportions of *SS* explained by **stimulus**, **conditions** and interaction are, respectively,

$$\frac{95}{1261}=7.53\% \quad \frac{12}{160}=7.50\% \quad \frac{35}{160}=21.88\%$$

12.3 EXAMPLE OF A WW DESIGN

12.3.1 Introduction

In a study of the effect of spectacle-wearing on recognition of faces, 10 randomly selected subjects were each shown 60 photographs of faces, 30 male and 30 female. Within each set of 30 photographs, half were of faces wearing spectacles and half were of faces not wearing spectacles.

The factors were:

spectacles (2 levels, within-subjects)
sex (2 levels, within-subjects)

The mean numbers of photographs correctly recognized in each combination of conditions are set out in Table 12.4. The layout of the means table and its interpretation for the WW design follow the same principles as for the BB and BW designs. The effects are displayed as bar charts in Fig. 12.3. From the graphs in Fig. 12.3 it is evident that, in this sample, wearing spectacles makes it harder to recognize a face. Also, male faces are easier to recognize than female faces, and there is an interaction. The interaction shows that it is mostly for females that spectacles make recognition more difficult. There is hardly any effect of spectacle-wearing on recognition for male faces.

Table 12.4 Mean numbers of photographs correctly recognized

	Spectacles	**No spectacles**	Means
Male faces	10.80	11.05	10.925
Female faces	8.00	11.10	9.550
Means	9.4	11.075	10.2375

An analysis of variance is required before statements can be made about the presence of effects in the population.

12.3.2 Analysis of variance for the WW design

The summary table for analysis of variance is given in Table 12.5.

The rules for the analysis of variance of the WW design will be dealt with in section 12.4.

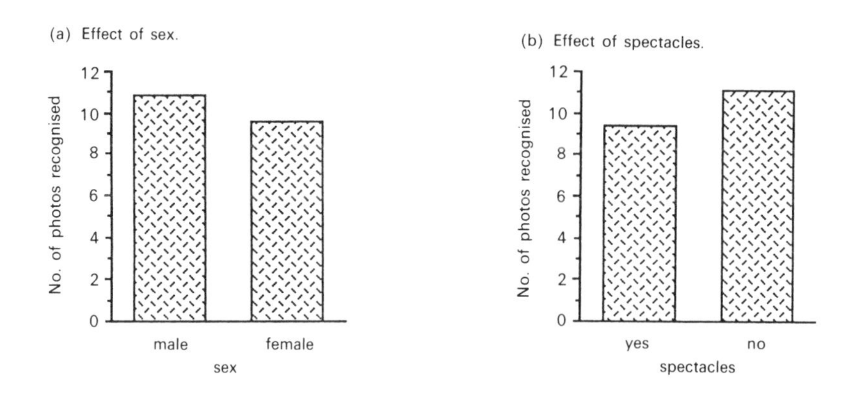

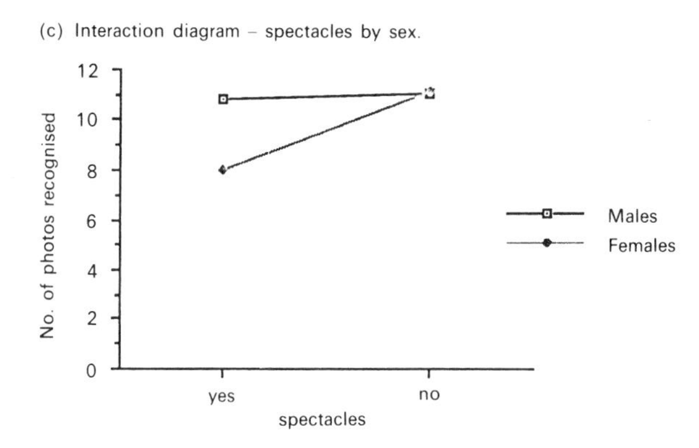

Fig. 12.3 WW design – mean numbers of photographs recognized.

12.3.3 Tests of significance for the WW design

The basic analysis of variance in Table 12.5 is obtained from most of the well-known statistical packages for computers (Appendix A). It makes available tests of hypothesis about the existence in the population of effects due to the two factors and their interaction. It is seen that both main effects and their interaction are statistically significant at the 0.05 level.

12.3.4 Venn diagram for the WW design

There should be different sections of the Venn diagram for the 'between' and 'within' portions of the total *SS* to indicate their independence. The between part is, however, of little interest, since it consists entirely of residual or unexplained variation. Figure 12.4 displays the Venn diagram for the WW example experiment.

Table 12.5 Analysis of variance for the WW design

Source	*df*	*Type*	*SS*	*MS*	MS_{error}	*F*	F_c
Spectacles	1	W	28.06	28.06	5.188	5.41	5.12
Sex	1	W	18.91	18.91	0.5404	34.99	5.12
Spectacles × Sex	1	W	20.31	20.31	0.4933	41.17	5.12
Subjects	9		91.28	not needed			
Subjects × Spectacles	9		46.69	5.188			
Subjects × Sex	9		4.864	0.5404			
Subjects × Spectacles × Sex	9		4.440	0.4933			
Total	39		214.554				

F_c is the 0.05 critical value

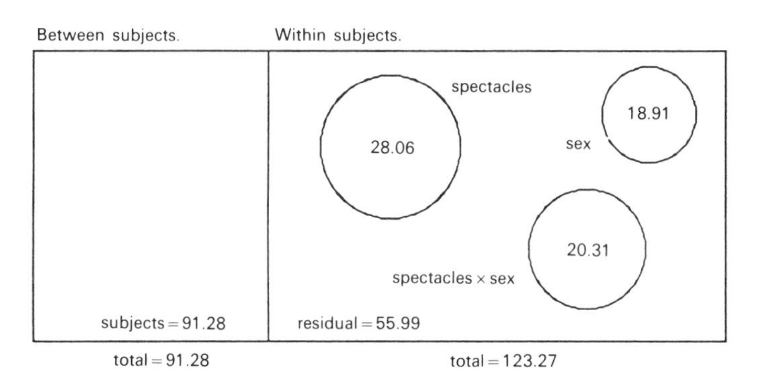

Fig. 12.4 Venn diagram for analysis of variance for WW design.

12.3.5 Sizes of effects for the WW design

The sizes of the effects of both factors and their interaction are obtained by expressing their *SS*s as proportions of the total *SS* within-subjects. The total *SS* within-subjects is the overall total *SS* minus the between-subjects *SS*. This comes to $214.554 - 91.28 = 123.274$.

Therefore the proportions of *SS* explained by **spectacles**, **sex** and interaction are, respectively,

$$\frac{28.06}{123.274} = 22.7\% \quad \frac{18.91}{123.274} = 15.3\% \quad \frac{20.31}{123.274} = 16.5\%$$

12.4 OVERVIEW OF RULES FOR THE ANOVA SUMMARY TABLE FOR DESIGNS BB, BW AND WW

12.4.1 Terms in the model (sources of variation) for all two-factor designs

The descriptions that follow are summarized in tabular form below. All three two-factor designs contain terms for each of the two factors and their interaction. In addition, they all contain a term which represents the variation between individual subjects. In the BB design this is called **within groups** or **subjects within groups**. In the design WW it is called **subjects**. In the design BW it is called either **subjects** or **subjects within groups**.

Finally, there are terms representing the interaction of **subjects** with any within-subject terms in the design. These all correspond to some form of reliability variation.

The complete picture, using an obvious notation, is:

BB	BW	WW
Factor B1	**Factor B**	**Factor W1**
Factor B2	**Factor W**	**Factor W2**
Interaction B1 × B2	**Interaction B × W**	**Interaction W1 × W2**
within groups	**subjects within groups**	**subjects**
	interaction:	**interactions:**
	subjects × W	**subjects × W1**
		subjects × W2
		subjects × W1 × W2

12.4.2 Degrees of freedom and *F*-test for all two-factor designs

The degrees of freedom (*df*) are obtained according to the rules set out here:

for the effect of a factor = (no. of levels − 1)
for interaction of factor 1 with factor 2 = (*df* factor 1) (*df* factor 2)
for subjects = (no. of groups of subjects) (no. of subjects per group − 1)
for interaction of subjects with a factor = (*df* subjects) (*df* factor)

Note that a useful check is provided by using the fact that the *df*s in the table must add up to the *df* of the total *SS*. This is easy to obtain as it is just one less than the number of measurements on the dependent variable obtained in the experiment.

Mean squares

Mean squares are formed for each term in the usual way by dividing the *SS* by the *df*:

$$MS = \frac{SS}{df}$$

F-test

The *F*-test is formed in the usual way by dividing the *MS* of the term to be tested by the appropriate error *MS*:

$$F = \frac{MS_{\text{term}}}{MS_{\text{error}}}$$

12.4.3 Choice of error *MS* for all two-factor designs

The denominator for calculation of the *F*-test is selected according to the following rules:

1. For the effect of a **between-subjects** factor, MS_{error} is MS_{subjects} (also known as *MS* within-groups).
2. For the effect of a **within-subjects** factor, MS_{error} is $MS_{\text{subjects} \times \text{factor}}$.
3. For the interaction of factor 1 with factor 2 the appropriate error term is set out in this table:

Factor 1	Factor 2	Error term
between	**between**	$MS_{\text{within-groups}}$
between	**within**	$MS_{\text{subjects} \times \text{factor 2}}$
within	**between**	$MS_{\text{subjects} \times \text{factor 1}}$
within	**within**	$MS_{\text{subjects} \times \text{factor 1} \times \text{factor 2}}$

Complete listing of error terms

For each of the three designs the factors and interaction in the top part of the table are linked by numbers (1), (2), etc. to the appropriate sources of error in the lower parts. These sources of error serve as the denominator or error *MS* for the *F*-test.

BB	BW	WW
Factor B1 (1)	**Factor B (1)**	**Factor W1 (2)**
Factor B2 (1)	**Factor W (2)**	**Factor W2 (3)**
Interaction B1 × B2(1)	**Interaction B × W (2)**	**Interaction W1 × W2(4)**
(1) **within groups**	(1) **subjects within groups**	(1) **subjects**
	interaction:	interactions:
	(2) **subjects × W**	(2) **subjects × W1**
		(3) **subjects × W2**
		(4) **subjects × W1 × W2**

12.5 TESTS OF SIGNIFICANCE FOR SIMPLE EFFECTS IN BW AND WW DESIGNS

12.5.1 Introduction

The concept of **simple effect** was introduced in section 2.3.1. The test of significance and a numerical example of simple effect in a BB design was introduced in section 6.4.3. The concept and technique apply equally in designs which include one or more within-subjects factors except for a detail of the test of significance. For some tests a **pooled MS_{error}** has to be calculated.

12.5.2 Calculation of *SS* of simple effect

The appropriate row or column of means, having been identified, is used to calculate the *SS* by the standard procedure. This procedure is described in sections 6.4.2 and 6.4.3 for between-subjects factors and in section 5.4.3 for within-subjects factors.

Examples of simple effects in BW design

(a) Suppose, in the above BW reaction time example in Table 12.2, that there is interest in the simple effect of **conditions** at the *rounded* level of **stimulus**. The appropriate row of means is:

(9.83 8.71)

This is transformed into a row of deviations which add to zero by subtracting from each mean the mean of the means in the row. This gives

(0.56 −0.56)

These deviations are squared and added and the result multiplied by n, where n is the number of individual measurements contributing to each mean in the row. Here n is 10, since 9.83 is the mean of 10 subjects' reaction times. This gives:

$$SS = 10(0.56^2 + (-0.56)^2) = 6.272$$

The degrees of freedom, following the usual rule, are one less than the number of means in the row. Here $df = 1$. Hence MS for the simple effect is $SS/df = 6.272$.

(b) Consider the simple effect of **stimulus** at the *control* level of **conditions**. The relevant column of means is

(9.83 10.73 8.91)

which as deviations is

(0.01 0.91 −0.91)

Thus

$$SS = 10(0.01^2 + 0.91^2 + (-0.91)^2) = 16.563$$

with two degrees of freedom. Hence the MS for this simple effect is $16.563/2 = 8.2815$.

For both examples (a) and (b) the identification of the appropriate MS_{error} for calculation of the F-ratio requires more care than in previous applications. The next section addresses this.

12.5.3 Identification of MS_{error} for the simple effect

MS_{error} depends on the design. The three two-factor designs are dealt with separately.

MS error in the BB design

All tests, including those of simple effects, use $MS_{\text{within-groups}}$ (which is also known as $MS_{\text{subjects within-groups}}$ or as MS_{subjects}).

MS error in the BW design

Factor W is within-subjects (repeated measures). Factor B is between-subjects (independent groups).

For a simple effect among the means of the within-subjects factor (at some particular level of factor B), use $MS_{\text{subjects} \times \text{W}}$ (i.e. the same error *MS* used in the *F*-test of factor W).

For a simple effect among the means of the between-subjects factor (at some particular level of factor W), use the specially constructed *MS* formed by pooling MS_{subjects} and $MS_{\text{subjects} \times \text{W}}$

$$MS_{\text{pooled}} = \frac{SS_{\text{subjects}} + SS_{\text{subjects} \times \text{W}}}{df_{\text{subjects}} + df_{\text{subjects} \times \text{W}}}$$

Examples from the BW design

(a) For the effect of **conditions** at the **rounded** level of **stimulus** it has already been shown that *MS* for the simple effect is 6.272. Since **conditions** is within-subjects, **subjects** × **W** forms the appropriate MS_{error}. Thus the test of significance is as follows:

$$F = 6.272/4.2 = 1.49$$

This has to exceed the critical *F* on 1 and 27 degrees of freedom. $F_c = 4.21$; do not reject H_0.

(b) For the simple effect of **stimulus** at the *control* level of **conditions**, it has already been shown, in section 12.5.2, that the *MS* is 8.2815. Since **stimulus** is between-subjects the **pooled** MS_{error} is required. The above formula gives:

$$MS_{\text{pooled}} = \frac{1166 + 113}{27 + 27} = 23.685$$

The degrees of freedom of MS_{pooled} can be taken, for a rough approximation, as the mean of the *df*s of the constituent *MS*s. In this case it is (27 + 27)/2, which is 27.

(A better approximation (Satterthwaite, 1946) is calculated from a formula given in Appendix D. It gives the *df* as 32.)

This leads to $F = 8.2815/23.685 = 0.35$, which has to exceed F_c with 2 and 27 degrees of freedom. $F_c = 3.35$; do not reject H_0.

MS error in the WW design

All tests on simple effects use pooled *MS* error terms.

Suppose the factors are known as W1 and W2. For the simple effect among the means of factor W1 (at some particular level of factor W2), use the *MS* formed by pooling $MS_{\text{subjects} \times \text{W1}}$ and $MS_{\text{subjects} \times \text{W1} \times \text{W2}}$:

$$MS_{\text{pooled}} = \frac{SS_{\text{subjects} \times \text{W1}} + SS_{\text{subjects} \times \text{W1} \times \text{W2}}}{df_{\text{subjects} \times \text{W1}} + df_{\text{subjects} \times \text{W1} \times \text{W2}}}$$

A rough approximation for the *df* for MS_{pooled} may be taken as the mean of $df_{\text{subjects} \times \text{W1}}$ and $df_{\text{subjects} \times \text{W2}}$. Alternatively, a better approximation is described in Appendix D.

12.5.4 Summary of procedure for tests of simple effects

The test is carried out by calculating

$$F = \frac{MS \text{ for simple effect}}{\text{appropriate } MS_{\text{error}}}$$

where *MS* for the simple effect almost always has to be calculated by hand (see section 6.4.3 for an example of calculation of this *MS* from the means which describe the effect). MS_{error} is either taken directly from the overall (omnibus) ANOVA or is a pooled *MS* based on terms from the overall ANOVA.

F, as calculated above, must be compared with F_c, the critical value of *F* based on the simple effect *df* and the error *df*. The simple effect *df* is just (no. of levels −1). The *df* of the error *MS* is taken from the row of the ANOVA table from which the error *MS* itself was taken. If the pooled *MS* is used, its *df* can be taken, for a rough approximation, to be the average of the two *df*s on which it is based (see appendix D for a better approximation).

The pro forma in section 12.6 should be copied and used as an aid to calculating the significance test of simple effects in two-factor designs.

12.6 CALCULATION PRO FORMA FOR SIMPLE EFFECTS IN TWO-FACTOR DESIGNS

1. Identify the set of means whose differences are to be tested.

   Means: _________ _________ _________ _________

2. Each of these means is obtained as the average of *n* measurements.

   Note $n =$ _________

3. Convert the set of means in step 1 into deviations by subtracting from each the overall mean of the set (if it is not already available calculate it by averaging the set of means by hand).

   Deviations: _________ _________ _________ _________

4. Calculate SS for the simple effect as in section 6.4.2.

   $SS = n$(sum of squares of deviations) = _________

5. Convert the SS into an MS by dividing by the appropriate degrees of freedom (df is number of means in the set less one).

   $MS =$ _________

6. Identify the appropriate MS_{error} from the overall ANOVA summary table. This follows the rules set out in section 12.5.3. Decide whether the design is BB, BW or WW.

   $MS_{\text{error}} =$ _________

7. Calculate F and hence test significance of the simple effect.

   $$F = \frac{MS}{MS_{\text{error}}} = \frac{\_\_\_\_\_\_\_\_\_}{\_\_\_\_\_\_\_\_\_} = \_\_\_\_\_\_\_\_\_$$

   df for numerator = _________ df for denominator = _________ (The df of MS_{error} either comes with it directly from the overall ANOVA table, or, if a pooled MS_{error} has been calculated, df is calculated as an average of the two relevant dfs for a rough approximation. See Appendix D for a better approximation.

   Obtain F_c from Appendix F.2.

   $F_c =$ _______

   If F exceeds F_c with the appropriate df you can decide to reject H_0 and conclude that the factor in question has an effect at the particular level of the other factor.

12.7 CONTRASTS AND COMPARISONS IN THE BW AND WW DESIGNS

Contrasts and comparisons may be tested among the levels of the factors in the BW and WW designs in the same way as in the one-factor designs, as discussed in Chapter 8.

The only area of possible difficulty is the identification of the appropriate error MS for the test of significance. The simple rule is that the same MS_{error} is used for testing a contrast among the levels of a factor as is used for testing the main effect of the factor. This is tabulated in Table 12.6. Note that $MS_{\text{subjects}\times\text{factor}}$ was referred to as $MS_{\text{reliability}}$ in Chapter 5 and section 8.2.6.

Table 12.6 MS_{error} for test of contrast in BW or WW design

	a priori	*a posteriori*
Between-subjects	MS_{subjects}	MS_{subjects}
Within-subjects	$MS_{\text{subjects}\times\text{factor}}$	N/A

As mentioned in Chapter 8, no *a posteriori* comparisons are available for within-subjects factors.

12.7.1 Example of BW design

Consider again the reaction time example of section 12.2.1. Suppose there was a requirement to test the contrast of *rounded* and *straight* levels of **stimulus**, considered together, with the *irregular* level. The linear contrast function would have the value

$$(-1)(9.27)+(-1)(12.04)+(+2)(9.48)=-2.35$$

and SS the value

$$\frac{(20)(-2.35)^2}{1^2+1^2+2^2}=18.41$$

The multiplier n has the value 20. This is because each mean, such as 9.27, is based on 20 measurements of reaction time.

The appropriate error MS is MS_{subjects}, since **stimulus** is a between-subjects factor and so tested using MS_{subjects} (also known as $MS_{\text{within-groups}}$). Hence $F=18.41/43.2=0.426$ with 1 and 27 degrees of freedom. F_c is 4.21; do not reject H_0.

For the *a posteriori* version of this test the Scheffé adjusted F would be $(2)F(2, 27)$, i.e. $(2)(3.35)=6.7$ at the 0.05 level of significance. Note that as in the single-factor design, the Scheffé adjusted F_c is just $(k-1)$ times the critical F used in the test of the main effect.

12.8 EXERCISES

12.1 Eight subjects were divided into two groups of four. Each group was administered a different drug treatment and then each subject undertook three learning tasks. Completion times (in seconds) were recorded and are shown in the table.

		Tasks				Main effect mean
	Subject	**T1**	**T2**	**T3**	Mean	
Drug A	1	32	30	17	26.33	
	2	34	26	22	27.33	
	3	28	24	19	23.67	
	4	32	27	21	26.67	
	Mean	31.50	26.75	19.75		26.00
Drug B	5	27	26	18	23.67	
	6	24	25	17	22.00	
	7	30	23	17	23.33	
	8	28	29	20	25.67	
	Mean	27.25	25.75	18.00		23.67
Main effect mean		29.375	26.250	18.875		

The sums of squares for the analysis of variance are:

Source	*SS*
Drugs	32.67
Tasks	465.08
Drugs × tasks	11.58
Subjects	43.95
Subjects × tasks	52.05
Total	605.33

(a) Identify the degrees of freedom, complete the ANOVA summary table and carry out the F-tests for main effects and interaction.
(b) Use appropriate graphs to display the main effects and their interaction, whether significant or not.
(c) Calculate the *SS* for the simple effect of task for Drug A and complete the F-test.
(d) Carry out an *a priori* comparison of Task 1 against Task 2.
(e) Briefly summarize the findings of the above analysis.

12.2 An experiment was carried out on movement control. Six subjects, all right-handed, having normal eyesight, male and aged 20 to 23 years were tested under all combinations of two experimental factors.

Subjects were required to draw lines of four different **lengths** (16 cm, 20 cm, 24 cm and 32 cm) and in eight different **directions** (0, 35, 90, 145, 180, 215, 270 and 325 degrees).

The dependent variable was the closest distance of the drawn line to a target point (**error distance**).

The statistical significance of the effects of factors **length** and **direction** and their interaction were investigated by analysis of variance.

The sums of squares (*SS* s) were:

Source	*SS*	*Source*	*SS*
Length (L)	5.70	**S × L**	13.59
Direction (D)	15.63	**S × D**	23.64
L × D	14.53	**S × L × D**	54.35
Subjects (S)	28.34		

(a) Complete all tests of significance.

(b) Given that the mean **error distance** in cm at the various **lengths** was:

Length	Mean **error distance**
16	1.27
20	1.34
24	1.40
32	1.72

(i) Carry out an *a priori* test for trend of increasing **error distance** with greater **length**.
(ii) Carry out an *a priori* test of the comparison of the mean **error distance** for the three shorter **lengths** (16, 20, 24 cm) with the longest **length** (32 cm).

(c) Give your view of the value of the selection criteria of the subjects.

12.3 Obtain the values of the terms in the model for the WW design based on the means in Table 12.4.

Three-factor designs

13.1 INTRODUCTION

Three-factor designs are a direct development of the two-factor designs that were dealt with in Chapters 6 and 12. They require of their users increased attention to organizing and presenting the results and minor extensions of the ideas of interaction and simple effects.

The treatment in this section differs from that in Chapter 6 in that no account is given either of the underlying logic of the tests of significance or of the formulae for hand calculation. The former follows from that of the one- and two-factor designs; the latter should be left to a computer package.

The researcher may choose a between- or within-subjects arrangement for each independent variable in a factorial design. For this reason the use of all combinations of between- and within-subjects factors is dealt with. There are four designs. They are referred to by the following shorthand:

- BBB — Three-factor independent groups design. All factors are between-subjects. A different group of subjects is exposed to each combination of levels of the three factors.
- BBW — Two between-subjects factors and one within-subjects factor.
- BWW — One between-subjects factor and two within-subjects factors.
- WWW — Three within-subjects factors. A single sample of subjects is taken through all combinations of levels of the three factors.

Three-factor designs are frequently required in professional research and increasingly in student projects now that computers are readily available to analyse the results. There are several reasons why they may be required in preference to simpler designs:

- Economies are achieved through having the individual subjects respond to the combined effects of several factors. This may avoid the need to carry out separate experiments for each factor.
- Information is obtained on interactions.
- One or more blocking factors can be included to control the conditions under which the other factors are investigated. This can improve the power or permit a reduced sample size. (section 9.3)

13.2 EXAMPLE OF BBB DESIGN

This example is taken from Maxwell and Delaney (1990).

13.2.1 Introduction

A study was carried out into the effects of three possible treatments, bio-feedback, drug therapy and diet therapy, on hypertension. (Hypertension is the condition characterized by clinically high levels of blood pressure.)

The treatment factors were:

Name	*Description of treatment*	*Levels*
drug	medication	drug X, drug Y, drug Z
biofeed	physiological feedback	present, absent
diet	special diet	present, absent

All 12 combinations of the three treatments were included in the design. This makes a $3 \times 2 \times 2$ design with 12 treatment combinations (known as cells).

Because the treatments were expected to have long-term effects it was necessary to use a different group of subjects for each combination of treatment conditions. This is equivalent to saying that it is an independent groups or BBB design. Seventy-two subjects were randomly sampled from the chosen population and allocated randomly, six to each combination of treatments.

The design compares the three drug treatments and compares presence with absence for two non-drug treatments. All subjects experience one of the three drugs. There is no *control group*, kept free of all active treatments. The design does not permit a test of **diet** or **biofeed** in the absence of medication. This could be provided if one of the drugs were a placebo. (A placebo is a control treatment which, to the subjects and workers participating in the experiment, is indistinguishable from the drugs with active ingredients.)

13.2.2 Organizing and presenting the raw data

The blood pressure measurements on the individual subjects are presented in Table 13.1 The data in Table 13.1 represent values of the dependent variable. The independent variables are the levels of the three factors **drug**, **biofeed** and **diet**.

For entry into a computer it is necessary to create accompanying values for the independent variables, as is done in Table 13.2. In the table the values 1 and 2 stand for *absent* and *present*, respectively, for both **diet** and **biofeed**. The values 1, 2 and 3 represent the levels X, Y and Z of the **drug** factor.

Table 13.1 Blood pressure data

Biofeed *and* **drug** *X*	**Biofeed** *and* **drug** *Y*	**Biofeed** *and* **drug** *Z*	**Drug** *X* *alone*	**Drug** *Y* *alone*	**Drug** *Z* *alone*
Diet absent					
170	186	180	173	189	202
175	194	187	194	194	228
165	201	199	197	217	190
180	215	170	190	206	206
160	219	204	176	199	224
158	209	194	198	195	204
Diet present					
161	164	162	164	171	205
173	166	184	190	173	199
157	159	183	169	196	170
152	182	156	164	199	160
181	187	180	176	180	179
190	174	173	175	203	179

Table 13.2 Data for BBB example arranged for computer entry

Diet	Biofeed	Drug	BP
1	2	1	170
1	2	1	175
1	2	1	165
⋮	⋮	⋮	⋮
2	1	3	160
2	1	3	179
2	1	3	179

13.2.3 Organizing and presenting the means

For designs with three or more factors, organizing and presenting the means demands careful attention. The straightforward approach (adopted by the SAS ANOVA program, Appendix A) is to list them separately for each main effect and interaction. This is reproduced in Table 13.3. In Table 13.3, means tables are set out in the order:

diet
biofeed
diet × **biofeed**
drug
diet × **drug**
biofeed × **drug**
diet × **biofeed** × **drug**

Table 13.3 also shows the number of blood pressure (BP) measurements averaged to give each mean. In the first row of the table the mean blood pressure of the 36 individuals in the *diet absent* condition is shown as 193.0.

Table 13.3 Mean blood pressures for the BBB example

Diet	Biofeed	Drug	*No.*	*Mean* **BP**
Absent			36	193.0
Present			36	176.0
	Present		36	179.2
	Absent		36	189.8
Absent	Present		18	187.0
Absent	Absent		18	199.0
Present	Present		18	171.3
Present	Absent		18	180.7
		Drug X	24	174.5
		Drug Y	24	190.8
		Drug Z	24	188.3
Absent		**Drug X**	12	178.0
Absent		**Drug Y**	12	202.0
Absent		**Drug Z**	12	199.0
Present		**Drug X**	12	171.0
Present		**Drug Y**	12	179.5
Present		**Drug Z**	12	177.5
	Present	**Drug X**	12	168.5
	Present	**Drug Y**	12	188.0
	Present	**Drug Z**	12	181.0
	Absent	**Drug X**	12	180.5
	Absent	**Drug Y**	12	193.5
	Absent	**Drug Z**	12	195.5
Absent	Present	**Drug X**	6	168.0
Absent	Present	**Drug Y**	6	204.0
Absent	Present	**Drug Z**	6	189.0
Absent	Absent	**Drug X**	6	188.0
Absent	Absent	**Drug Y**	6	200.0
Absent	Absent	**Drug Z**	6	209.0
Present	Present	**Drug X**	6	169.0
Present	Present	**Drug Y**	6	172.0
Present	Present	**Drug Z**	6	173.0
Present	Absent	**Drug X**	6	173.0
Present	Absent	**Drug Y**	6	187.0
Present	Absent	**Drug Z**	6	182.0

The more compact way to display the means involves setting them out in a series of two-way tables of rows and columns with row and column means appended. Table 13.4 is an example of this.

Table 13.4 displays each set of main effect means twice and omits the three-way interaction, but is easy to use. Each of its three two-way tables is said to be 'collapsed over the levels' of one factor. For example, the first is the **biofeed** by **diet** table; it displays means obtained by *collapsing over the levels* of **drug**. In other words, factor **drug** has been ignored for the purposes of this two-way table. The other two tables are the result of collapsing over the levels of factors **biofeed** and **diet**.

The three-way interaction is conveniently displayed in a two-way table, as shown in Table 13.5.

Table 13.4 Mean blood pressures displayed in two-way interaction tables for the BBB example

		Biofeed		
		Present	Absent	Means
Diet	Absent	187.0	199.0	193.0
	Present	171.3	180.7	176.0
	Means	179.2	189.8	

		Drug			
		Drug X	**Drug Y**	**Drug Z**	Means
Diet	Absent	178.0	202.0	199.0	193.0
	Present	171.0	179.5	177.5	176.0
	Means	174.5	190.8	188.3	

		Drug			
		Drug X	**Drug Y**	**Drug Z**	Means
Biofeed	Present	168.5	188.0	181.0	179.2
	Absent	180.5	193.5	195.5	189.8
	Means	174.5	190.8	188.3	

Table 13.5 Three-way interaction mean blood pressures displayed for the BBB example

	Biofeed	**Drug X**	**Drug Y**	**Drug Z**
Diet absent	Present	168.0	204.0	189.0
	Absent	188.0	200.0	209.0
Diet present	Present	169.0	172.0	173.0
	Absent	173.0	187.0	182.0

The means in Table 13.5 describe the **three-way interaction** or **three-factor interaction**. For example, the means

168.0 204.0 189.0
188.0 200.0 209.0

describe the interaction between **biofeed** and **drug** in the *absence* of diet, whereas the following means:

169.0	172.0	173.0
173.0	187.0	182.0

describe the interaction between **biofeed** and **drug** in the *presence* of diet. See section 13.2.5 for the interpretation of this.

If there is no interest in the three-way interaction then the complete set of tables of two-way interactions with their marginal means, as shown in Table 13.4, is normally the preferred presentation.

13.2.4 The means in the form of an additive model

For consistency with the treatment of the two-factor design in Chapters 6 and 12 there now follows a complete expression of the means in the form of the values of the components of the additive model. This differs from the earlier accounts in sections 6.7 and 12.2 in that it includes a term for the three-factor interaction. (In order to reduce the effects of rounding errors, all the deviations in the model are expressed to two places of decimals.)

$$\text{score} = \underset{\text{mean}}{\text{overall}} + \textbf{biofeed} + \textbf{drug} + \textbf{diet} +$$

$$= 184.5 + \left\{\begin{matrix} -5.33 \\ +5.33 \end{matrix}\right\} + \left\{\begin{matrix} -10.00 \\ +\ 6.25 \\ +\ 3.75 \end{matrix}\right\} + \left\{\begin{matrix} +8.50 \\ -8.50 \end{matrix}\right\} +$$

$$\left\{\begin{matrix} \text{present} \\ \text{absent} \end{matrix}\right\} \qquad \left\{\begin{matrix} \text{X} \\ \text{Y} \\ \text{Z} \end{matrix}\right\} \qquad \left\{\begin{matrix} \text{absent} \\ \text{present} \end{matrix}\right\}$$

$$\textbf{biofeed} \times \textbf{drug} + \textbf{biofeed} \times \textbf{diet} + \textbf{diet} \times \textbf{drug} +$$

$$\left\{\begin{matrix} -0.67 & +2.58 & -1.91 \\ +0.67 & -2.58 & +1.91 \end{matrix}\right\} + \left\{\begin{matrix} -0.67 & +0.67 \\ +0.67 & -0.67 \end{matrix}\right\} + \left\{\begin{matrix} -5.00 & +2.75 & +2.25 \\ +5.00 & -2.75 & -2.25 \end{matrix}\right\} +$$

$$\left\{\begin{matrix} \text{present} \\ \text{absent} \end{matrix}\right\} \{\text{X} \quad \text{Y} \quad \text{Z}\} \qquad \left\{\begin{matrix} \text{present} \\ \text{absent} \end{matrix}\right\} \{\text{abs.} \quad \text{pres.}\} \qquad \left\{\begin{matrix} \text{absent} \\ \text{present} \end{matrix}\right\} \{\text{X} \quad \text{Y} \quad \text{Z}\}$$

$$\textbf{biofeed} \times \textbf{drug} \times \textbf{diet}$$

$$\textbf{diet}\text{ absent} \qquad\qquad \textbf{diet}\text{ present}$$

$$\left\{\begin{matrix} -3.33 & +5.42 & -2.09 \\ +3.33 & -5.42 & +2.09 \end{matrix}\right\} \qquad \left\{\begin{matrix} +3.33 & -5.42 & +2.09 \\ -3.33 & +5.42 & -2.09 \end{matrix}\right\}$$

$$\left\{\begin{matrix} \text{present} \\ \text{absent} \end{matrix}\right\} \{\text{X} \quad \text{Y} \quad \text{Z}\}$$

Hence the expected score for a person with **biofeed** {present}, **drug** {Y} and **diet** {absent} is

$$184.5+(-5.33)+(+6.25)+(+8.50)+(+2.58)+(-0.67)+(+2.75)+(+5.42)=204.0$$

This corresponds to the observed value in Tables 13.3 and 13.5.

13.2.5 Analysis of variance and interpretation

Analysis of variance for BBB example

The sums of squares for main effects and interactions are obtained from the deviations of the means or by using formulae similar to those in section 6.6.1. This is normally done by a statistical computer program (see Appendix A). The sums of squares within groups is obtained by pooling the sums of squares from every group of subjects. This follows the principles established for one- and two-factor designs.

The resulting analysis of variance summary table is set out in Table 13.6.

Table 13.6 ANOVA summary table for the BBB example

Source	*df*	*SS*	*MS*	*F*	*Significance*
Diet	1	5202	5202	33.204	**
Biofeed	1	2048	2048	13.072	**
Drug	2	3675	1837.5	11.729	**
Diet × Biofeed	1	32	32	0.204	NS
Diet × Drug	2	903	451.5	2.882	NS
Biofeed × Drug	2	259	129.5	0.827	NS
Diet × Biofeed × Drug	2	1075	537.5	3.431	*
Within groups	60	9400	156.667		
Total	71	22 594			

* is significant at the 0.05 level
** is significant at the 0.01 level
NS is not significant.

The critical values of F that are relevant to the tests of significance are those based on (1, 60) and (2, 60) degrees of freedom. These are:

at the 0.05 level: 4.00 and 3.15
at the 0.01 level: 7.08 and 4.98

The conclusion is that all three main effects are significant at the 0.01 level, none of the two-way interactions are significant and the three-way interaction is significant at the 0.05 level.

Interactions in the BBB example

The existence of a three-way interaction greatly complicates the interpretation. It implies that two-way interactions cannot be taken at face value.

Consider the interaction **diet** × **biofeed**. It is not significant, and the interaction diagram in Fig. 13.1 shows the parallel lines characteristic of no

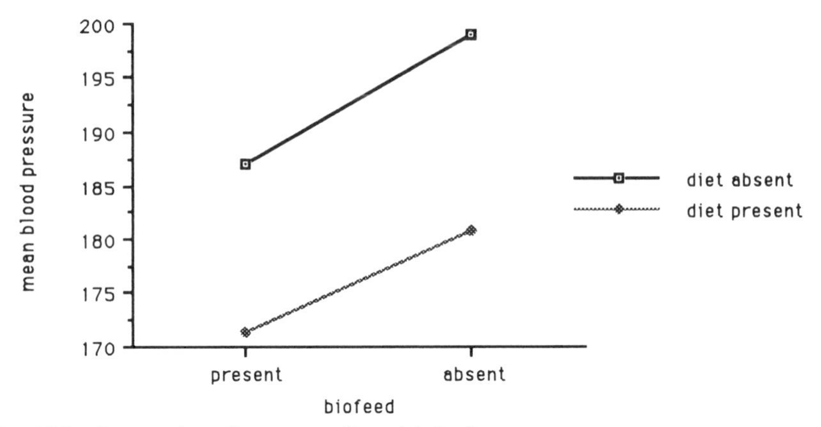

Fig. 13.1 Interaction diagram – diet × biofeed

interaction. This implies that the benefit of bio-feedback is the same whether or not the diet is taken.

However, the meaning of three-way interaction is that the interaction between any two of the factors is different at the different levels of the third factor.

Consider the **diet** × **biofeed** interaction diagrams obtained separately for the data from each of the **drugs X**, **Y** and **Z**. These are shown in Fig. 13.2. They are known as the *simple interaction effects* of **diet** with **biofeed**.

There is a marked contrast between the diagrams for **drugs X** and **Y**. For **drug X** the **diet** effect is greater at the **biofeed** absent level, whereas for **drug Y** the **diet** effect is greater at the **biofeed** present level. This implies that no clear statement can be made about the overall **diet** × **biofeed** interaction. Rather, the three simple interaction effects need to be reported and tested for statistical significance. The technique for this is dealt with in section 13.5.4.

The same discussion could have taken place concerning either the **diet** × **drug** or the **biofeed** × **drug** interactions. The existence of a significant three-way interaction implies that none of the overall two-way interactions is a coherent concept. All of them should be dealt with as separate simple interaction effects.

Main and simple effects in the BBB example

Significant interactions undermine the coherence of the concept of the main effect of a factor. Here the main effects are very strong. **Diet**, **biofeed** and **drug**, respectively, account for 23%, 9% and 16% of the total *SS* of 22 594, whereas the interactions altogether only account for 10%.

However, some of the simple effects may be interesting. Note that there are two levels of simple effect in the three-factor design: *first-order simple effects*, in which one of the other factors is fixed at a particular level, and

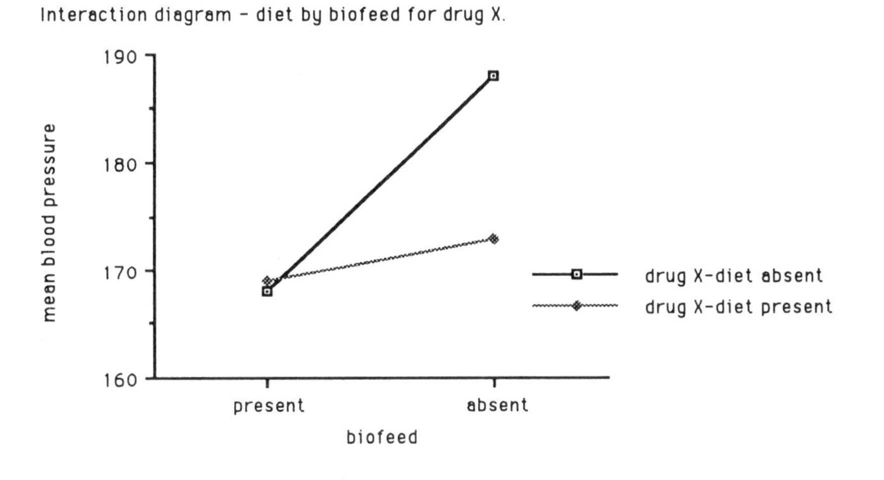

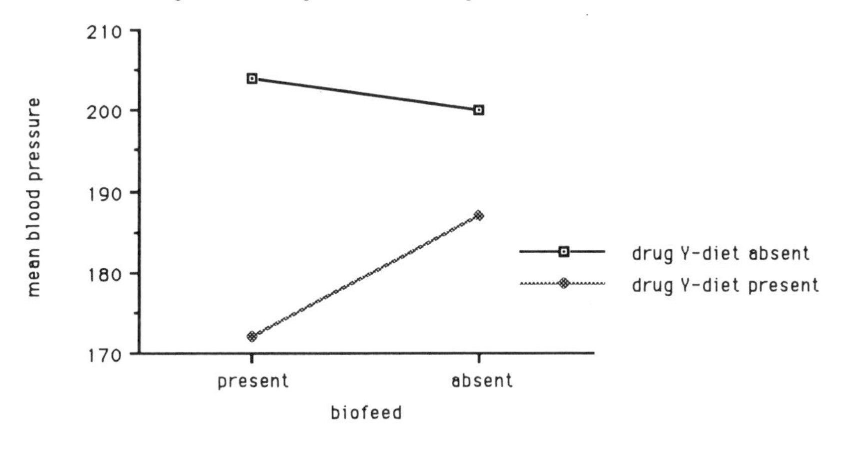

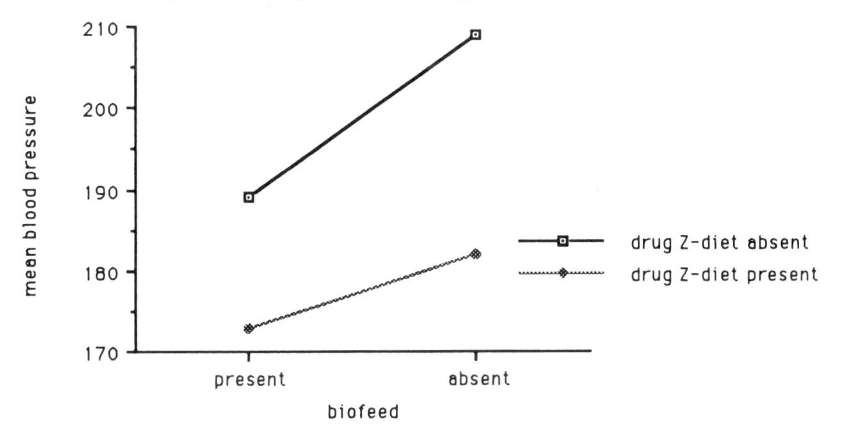

Fig. 13.2 Interaction diagrams – **diet** × **biofeed** for drugs X, Y and Z

second-order simple effects, in which both of the other factors are fixed at particular levels.

For example, consider the simple effect of **drug** for **diet** *absent*. This is a first-order simple effect represented by the means 178.0, 202.0 and 199.0. Consider the simple effect of **drug** for **diet** *absent* and **biofeed** *present*. This is a second-order simple effect represented by the means 168.0, 204.0 and 189.0. The tests of significance of simple effects are dealt with in section 13.5.2.

A complete analysis of this data can be found in Maxwell and Delaney (1990, pp. 325–38).

13.3 EXAMPLE OF A BBW DESIGN

This example is loosely based on Shore (1958) and given consideration in Winer (1991).

An experimenter was interested in evaluating the effect of anxiety and muscular tension on a learning task.

Subjects who scored extremely low on a scale measuring manifest anxiety are assigned to *level 1* and those who scored extremely high are assigned to *level 2* of the **anxiety** factor.

Other subjects were excluded from the experiment.

The **tension** factor is defined by pressure required to be exerted on a dynamometer. One half of the low-anxiety subjects are assigned at random to the low-tension condition; the other half are assigned to the high-tension condition. The high-anxiety subjects are divided in a similar manner. Subjects carry out the learning task once a day for four successive days. The dependent variable is the number of errors in the tests which take place at the conclusion of each day's learning task.

The results are presented in Table 13.7. Note that the first two columns of Table 13.7 represent levels of the between-subjects factors **anxiety** and

Table 13.7 Data for the BBW example

Anxiety	**Tension**	*Subject no.*	**Occasion**			
			1 d.v.	**2** d.v.	**3** d.v.	**4** d.v.
1	1	1	18	14	12	6
1	1	2	19	12	8	4
1	1	3	14	10	6	2
1	2	4	16	12	10	4
1	2	5	12	8	6	2
1	2	6	18	10	5	1
2	1	7	16	10	8	4
2	1	8	18	8	4	1
2	1	9	16	12	6	2
2	2	10	19	16	10	8
2	2	11	16	14	10	9
2	2	12	16	12	8	8

Table 13.8 Means for interactions for the BBW example

		Occasion				
		1	**2**	**3**	**4**	Means
Anxiety	**Low**	16.17	11.00	7.83	3.17	9.54
	High	16.83	12.00	7.67	5.33	10.46
	Means	16.50	11.50	7.75	4.25	

		Occasion				
		1	**2**	**3**	**4**	Means
Tension	**Low**	16.83	11.00	7.33	3.17	9.58
	High	16.17	12.00	8.17	5.33	10.42
	Means	16.50	11.50	7.75	4.25	

		Anxiety		
		Low	**High**	Means
Tension	**Low**	10.42	8.67	9.54
	High	8.75	12.17	10.46
	Means	9.58	10.42	

Table 13.9 Summary of analysis of variance for the BBW example

Source	*df*	*SS*	*MS*	MS_{error}	*F*	*p*
Anxiety × Tension	1	80.1	80.1	10.3	7.77	0.0237
Anxiety	1	10.1	10.1	10.3	0.98	0.3517
Tension	1	8.33	8.33	10.3	0.81	0.3949
Subjects	8	82.5	10.3			
Occasion	3	991.5	330.5	2.17	152.30	0.0001
Occasion × Anxiety × Tension	3	12.8	4.25	2.17	1.96	0.1477
Occasion × Anxiety	3	8.42	2.81	2.17	1.29	0.3003
Occasion × Tension	3	12.2	4.06	2.17	1.87	0.1624
Occasion × Subjects	24	52.2	2.17			
Total	47					

tension. The within-subjects factor, **occasion**, is not represented in the same way. Rather, its four levels appear implicitly in the table, each represented by a column of scores of the dependent variable. Particular attention is required when such data is entered into a computer.

The means are set out in Table 13.8. The three-way interaction is omitted.

The analysis of variance summary is set out in Table 13.9. Note that the only statistically significant effects are the main effect of **occasion** and the interaction of **anxiety** with **tension**.

The significant effects are represented graphically in Figs 13.3 and 13.4. The absence of a statistically significant three-way interaction implies that the two-way interaction of **anxiety** × **tension** in Fig. 13.3 may be assumed to apply to each of the four occasions.

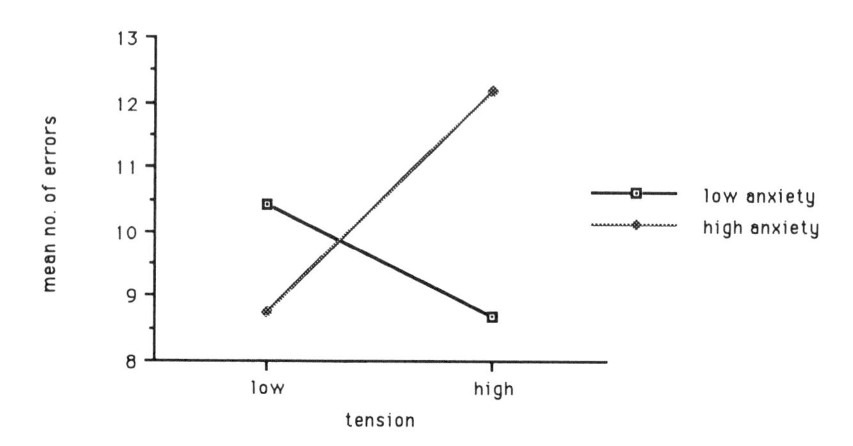

Fig. 13.3 Interaction diagram – **anxiety** × **tension**.

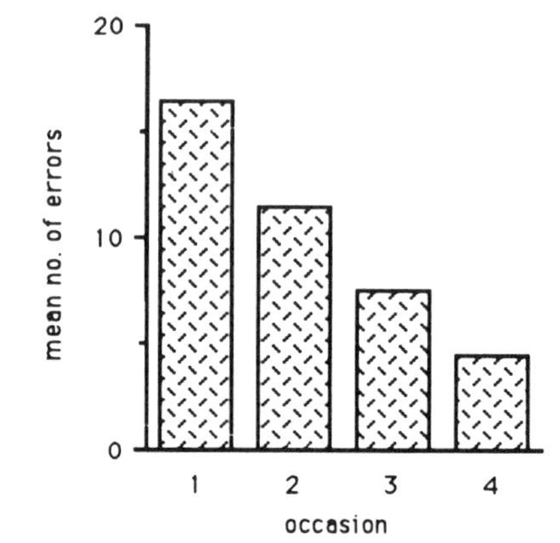

Fig. 13.4 Bar chart for main effect of **occasion**.

13.4 EXAMPLE OF A BWW DESIGN

As part of a research programme into the nature of dyslexia an experiment was carried out into the effect of speed of presentation of numbers on memory. Twelve dyslexic children and 12 normal children matched in pairs for mental age and digit memory ability were presented with three-digit numbers at four different speeds. The time intervals between numbers were 0.38, 1.00, 3.00 and 10.00 seconds.

Each subject was shown eight three-digit numbers at each speed. The order of presentation of the speeds was randomized. Mean error scores were obtained on each digit position at each speed. These 12 mean error scores for each child were the units for analysis. There were 288 such units in the analysis.

The design is a $2 \times 3 \times 4$ with repeated measures on two factors. The factors are:

Factor	*Description*	*Type*	*Levels*
group	type of child	between	2
digit	position of digit in number	within	3
speed	speed of presentation	within	4

The mean error scores at the various combinations of levels of the three factors are set out in Table 13.10, the means which describe the three-way interaction are set out in Table 13.11, and the summary of the analysis of variance is set out in Table 13.12. For comments on these results see section 13.5.

Table 13.10 Mean error scores displayed in two-way interaction tables for the BWW example

$n=36$ per cell		**Speed**				Means
		1	**2**	**3**	**4**	
Group	**Dyslexic**	2.53	2.37	1.97	1.67	2.13
	Normal	1.87	1.30	1.10	1.40	1.42
	Means	2.20	1.83	1.53	1.53	

$n=24$ per cell		**Speed**				Means
		1	**2**	**3**	**4**	
Digit	**First**	0.85	0.80	1.10	0.65	0.85
	Second	2.00	1.60	1.30	1.70	1.65
	Third	3.75	3.10	2.20	2.25	2.83
	Means	2.20	1.83	1.53	1.53	

$n=48$ per cell		**Digit**			Means
		First	**Second**	**Third**	
Group	**Dyslexic**	1.20	2.10	3.10	2.13
	Normal	0.50	1.20	2.55	1.42
	Means	0.85	1.65	2.83	

Table 13.11 Three-way interaction mean error scores displayed in a matrix for the BWW example

$n=12$ per cell		**Speed**			
Digit	**Group**	**1**	**2**	**3**	**4**
First	**Dyslexic**	1.00	1.40	1.60	0.80
	Normal	0.70	0.20	0.60	0.50
Second	**Dyslexic**	2.90	2.10	1.50	1.90
	Normal	1.10	1.10	1.10	1.50
Third	**Dyslexic**	3.70	3.60	2.80	2.30
	Normal	3.80	2.60	1.60	2.20

Table 13.12 Analysis of variance summary for the BWW example

Source	*df*	*SS*	*MS*	*Error term*	*F*
Group	1	36.98	36.98	**subjects**	17.7
Speed	3	21.66	7.22	**subjects × speed**	4.7
Digit	2	190.48	95.24	**subjects × digit**	104.7
Group × Speed	3	6.30	2.10	**subjects × speed**	1.4
Group × Digit	2	1.48	0.74	**subjects × digit**	0.8
Speed × Digit	6	26.52	4.42	**subjects × speed × digit**	2.0
Group × Speed × Digit	6	13.08	2.18	**subjects × speed × digit**	1.0
Subjects	22	45.98	2.09		
Subjects × Speed	66	101.64	1.54		
Subjects × Digit	44	40.04	0.91		
Subjects × Speed × Digit	132	293.04	2.22		
Total	287				

13.5 SUMMARY OF RULES FOR ANALYSIS OF BBB, BBW, BWW AND WWW DESIGNS

13.5.1 Introduction

The following is an account of main, simple and interaction effects in three-factor balanced designs, in which any of the factors may be within-subjects (repeated measures).

All examples are from the BWW memory-for-digits example in section 13.4.

13.5.2 Main effects

Definition

The main effect is a comparison among the mean scores at the various levels of a factor. Scores are averaged across all levels of other factors. In other words, the other factors are collapsed out.

For example: the main effect of digit position (**digit**) is the comparison among the means, each based on 96 measurements, of the scores obtained in the first, second and third digit position in the number.

Numerical representation

1. It is represented as a set of means, e.g. **digit** effect is 0.85, 1.65, 2.83.
2. It can also be represented as a set of deviations from the overall mean, 1.775, (t_1, t_2, t_3, ...) which add to zero, e.g. (−0.925, −0.125, 1.055).

Graphical representation

This is a bar chart of means, as shown in Fig. 13.5.

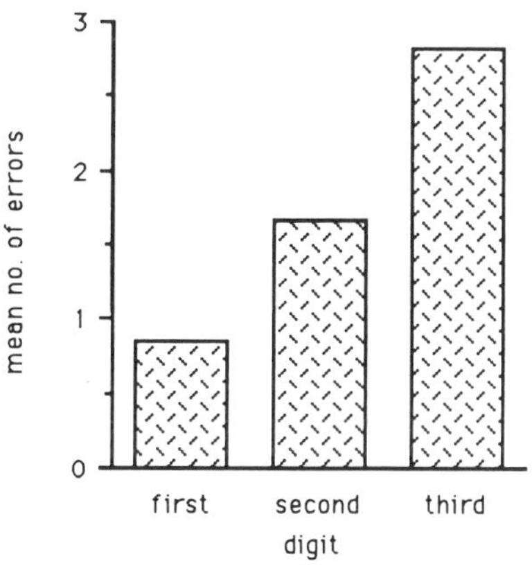

Fig. 13.5 Bar chart for digit main effect.

Sums of squares (SS)

The *SS* is defined in the usual way as

$$n(t_1^2+t_2^2+t_3^2+\cdots)$$

where n is the number of measurements at each level of the factor and t_1, t_2, etc. are the deviations representing the effect. For example, *SS* for **digit** $=96(0.925^2+0.125^2+1.055^2)=190.49$, as in the summary in Table 13.12 (subject to rounding error).

Size of effect

In absolute terms, the size of the effect of a factor is represented by the *SS* of the factor, although the *SS* also includes an element of sampling fluctuation. It can also, and more usefully, be expressed as a proportion of the total *SS* between-subjects or of the total *SS* within-subjects, whichever is appropriate.

The between-subjects sources of variance consist of the between-subjects factors, the interactions of the between-subjects factors with one-another and 'subjects' itself. 'Subjects' is also known as 'subjects within groups' or just as 'within groups' variation.

The within-subjects sources of variance consist of the within-subjects factors and the interaction of any factor with a within-subjects factor. For example,

SS for **digit** = 190.48

total SS within-subjects $= 21.66 + 190.48 + 6.30 + 1.48 + 26.52 + 13.08 + 101.64 + 40.04 + 293.04 = 694.24$

Hence **digit** accounts for 27% of the within-subjects variation.

Test of significance

This is formed by dividing the MS of the factor by the appropriate MS_{error}.

$$F = \frac{MS_{\text{factor}}}{MS_{\text{error}}}$$

In BBB all effects use $MS_{\text{within-groups}}$ as MS_{error}.

The other designs use MS_{subjects} (also known as $MS_{\text{subjects-within-groups}}$) as MS_{error} for between-subjects factors and $MS_{\text{factor} \times \text{subjects}}$ as MS_{error} for within-subjects factors.

For example: test of **digit** using $MS_{\text{subjects} \times \text{digit}}$ as MS_{error}:

$$F = \frac{95.24}{0.91} = 104.66$$

on (2, 44) degrees of freedom as in Table 13.12.

13.5.3 Simple effects

Definition

Simple effects are conceptually the same as the main effect, except that the comparison is among the mean scores of the levels of a factor at a single level of one of the other factors, for a first-order simple effect, or at single levels of both the other factors for a second-order simple effect. An alternative name for the simple effect could be 'partial main effect'.

For example, the simple effect of **digit** for the first level of **speed** will be based on means of the 24 scores for each digit position. This is a first-order simple effect and can be identified in the first column of the digit × speed part of Table 13.10. The simple effect of **digit** for the first level of **speed** for the dyslexic group is based on the means of the 12 scores for each digit position, at the first level of **speed** for the dyslexic group. This is a second-order simple effect and can be identified in the first column of Table 13.11.

Numerical representation

1. The simple effect is represented as a set of means, e.g. the simple effect of **digit** for the first level of **speed** as means is (0.85 2.00 3.75). The

simple effect of **digit** for the first level of **speed** for the dyslexic group as means is (1.00 2.90 3.70).

2. The simple effect may also be represented as a set of deviations from the appropriate overall mean. For example, the simple effect of **digit** for the first level of **speed** as deviations is (-1.35 -0.20 1.55) (note that $n=24$). The simple effect of **digit** for the first level of **speed** for the dyslexic group as deviations is (-1.53 0.37 1.17) (note that $n=12$).

Graphical representation

The graphical representation is a bar chart of means; similar to that for the main effect.

Sums of squares

The *SS* is obtained from the deviations by the usual formula. For example, the simple effect of **digit** for the first level of **speed** is

$$SS=24((-1.35)^2+(-0.20)^2+1.55^2)$$
$$=102.36$$

The simple effect of **digit** for the first level of **speed** for the dyslexic group is

$$SS=12((-1.53)^2+0.37^2+1.17^2)$$
$$=46.16$$

The multipliers are 24 and 12 for the first- and second-order simple effects, respectively, because they are the numbers of measurements on which are based the deviations.

Size of effect

This follows the same rule as for the main effect. That is, it can be expressed as the absolute value of *SS* or as a proportion of the appropriate total *SS*.

Test of significance

F is calculated as the *MS* of the simple effect divided by the MS_{error}. The simple effect *MS* is just the *SS* divided by *df*. *df* is one less than the number of levels of the factor. The MS_{error} is the $MS_{\text{within-groups}}$ in the BBB design.

For other designs the procedures are fairly complicated. For these the reader is referred to Winer *et al.* (1991, pp. 529–31, 535–7 and 550–1).

13.5.4 Interaction effect

Definition

The interaction between two factors is identical to the concept introduced for the two-factor design in Chapters 6 and 12.

By collapsing over the levels of the third factor the design reduces to a two-factor design. The interaction thus defined has been discussed at length in Chapters 6 and 12.

For example, the **digit** by **speed** interaction is based on comparisons among the 12 means in the **digit** by **speed** means table. These means are each based on the scores of all 24 children under each combination of digit position and speed of presentation.

The interaction is the comparison among the simple effects of **digit** at the various levels of **speed**. If all the simple effects of **digit** are the same regardless of the level of **speed**, then there is no interaction.

Numerical representation

1. As a set of means:

		Speed			
		1	**2**	**3**	**4**
Digit	first	0.85	0.80	1.10	0.65
	second	2.00	1.60	1.30	1.70
	third	3.75	3.10	2.20	2.25

2. As a set of deviations:

$$\begin{matrix} T_{11} & T_{12} & T_{13} & T_{14} \\ T_{21} & T_{22} & T_{23} & T_{24} \\ T_{31} & T_{32} & T_{33} & T_{34} \end{matrix}$$

		Speed			
		1	**2**	**3**	**4**
Digit	first	−0.42	−0.10	0.50	0.05
	second	−0.07	−0.10	−0.10	0.30
	third	0.50	0.22	−0.38	−0.33

A method of calculating the deviations is described in section 6.4.4. An easier method is described in section 13.5.5.

Graphical representation

This is a plot of the appropriate means as previously illustrated. It is called an interaction diagram, or profile diagram. The scale of the vertical axis is the mean score. For example, the **digit** × **speed** plot is shown in Fig. 13.6.

Sums of squares

The SS is defined to be $n(t_{11}^2 + t_{12}^2 + t_{13}^2 + \cdots + t_{22}^2 + \cdots)$, where n is the number of measurements at each combination of levels of the two factors.

For example, for **digit** × **speed**,

$$\begin{aligned} SS &= 24(0.42^2 + 0.10^2 + 0.50^2 + 0.05^2 + 0.07^2 + 0.10^2 + 0.10^2 + 0.30^2 + \\ &\quad 0.50^2 + 0.22^2 + 0.38^2 + 0.33^2) \\ &= 26.532 \end{aligned}$$

(subject to rounding errors) as in the ANOVA summary in Table 13.12.

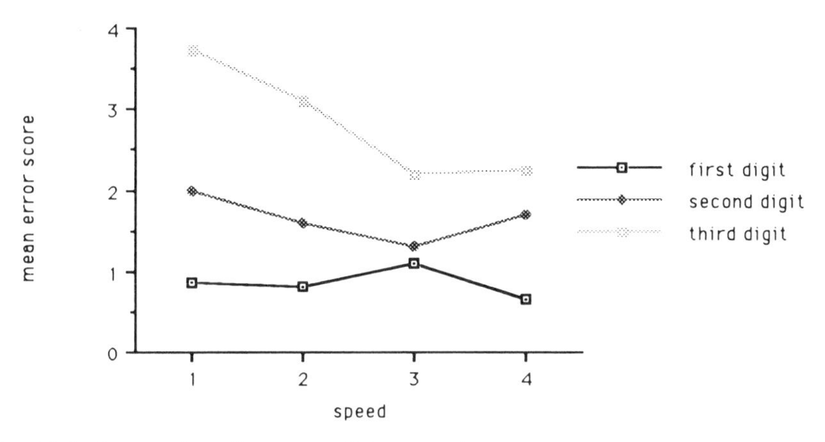

Fig. 13.6 Interaction diagram – **digit** × **speed**.

Size of effect

In absolute terms this is given by the value of the *SS*. It can also be expressed as a proportion of the total *SS* between-subjects or of the total *SS* within-subjects, whichever is appropriate. Only if both component factors of the interaction are between-subjects is it appropriate to express the *SS* as a proportion of the between-subjects *SS*. If either or both factors are within-subjects the *SS* should be expressed as a proportion of the total within-subjects *SS*.

For example, the **digit** × **speed** effect accounts for 26.532/694.24 = 3.82% of the total within-subjects variation.

Test of significance

The *F* value is formed by dividing the *MS* of the interaction by the appropriate MS_{error}. Suppose we are interested in testing the A × B interaction. The correct error term depends on whether factor A or B are within- or between-subjects. The rule is set out in this table:

A	B	*Error term*
between	between	$MS_{subjects}$
between	within	$MS_{subjects \times B}$
within	between	$MS_{subjects \times A}$
within	within	$MS_{subjects \times A \times B}$

The *df* for the interaction is the product of the *df*s of the two component factors, e.g. for the test of **digit** × **speed**, *df* = (2) (3) = 6.

$$MS_{digit \times speed} = 26.532/6 = 4.422$$

$$F = \frac{MS_{digit \times speed}}{MS_{subjects \times digit \times speed}} = \frac{4.422}{2.22} = 1.99$$

The degrees of freedom are (6, 132) for the test. This leads to a critical F of 2.10. H_0 cannot be rejected.

13.5.5 Simple interaction effect

Definition

The simple interaction between two factors is the interaction between them at a single level of the third factor. For example, the interaction **digit** × **speed** for dyslexic children only is a comparison among the mean scores for the various combinations of levels of **digit** and **speed**, but only using data from the dyslexic children.

Numerical representation

1. As cell means:

Dyslexics only		Speed			
		1	**2**	**3**	**4**
Digit	first	1.00	1.40	1.60	0.80
	second	2.90	2.10	1.50	1.90
	third	3.70	3.60	2.80	2.30

2. As deviations.
   The procedure for obtaining the deviations appropriate to an interaction was described in section 6.4.4. However, a faster method is illustrated for the above example and described in steps 1 to 3 as follows.
   Step 1 Set out an empty matrix with the row and column means and overall mean appropriate to the simple interaction.

Dyslexics only		Speed				Row
		1	**2**	**3**	**4**	means
Digit	first					1.20
	second					2.10
	third					3.10
Column means		2.53	2.37	1.97	1.67	2.13

Step 2 Fill in the expected values of the cell means based on the values of the row and column means according to the following rule:

$$\text{expected value} = \text{row mean} + \text{column mean} - \text{overall mean}$$

For example, the expected value for the first **digit** position and **speed** *level 1* is

$$1.20 + 2.53 - 2.13 = 1.60$$

The expected value for the second **digit** position and **speed** level 1 is

$$2.10 + 2.53 - 2.13 = 2.50$$

The complete array of expected values has to be obtained. The result is as follows:

Dyslexics only		Speed			
		1	**2**	**3**	**4**
Digit	first	1.60	1.44	1.04	0.74
	second	2.50	2.34	1.94	1.64
	third	3.50	3.34	2.94	2.64

Step 3 Subtract the expected values from the corresponding observed values to obtain the deviations.

For example, the deviation for the first **digit** position and **speed** level 1 is

$$1.00 - 1.60 = -0.60$$

The deviation for the second **digit** position and **speed** level 1 is

$$2.90 - 2.50 = 0.40$$

The complete array of deviations follows:

Dyslexics only		Speed			
		1	**2**	**3**	**4**
Digit	first	−0.60	−0.04	0.56	0.06
	second	0.40	−0.24	−0.44	0.26
	third	0.20	0.26	−0.14	−0.34

Graphical representation

Plot an interaction diagram as for the main interaction effect.

Sum of squares

This is the same as for the main interaction effect. For example, for **digit** × **speed** for dyslexics only:

$$\begin{aligned} SS &= 12((-0.60)^2 + (-0.04)^2 + \cdots + 0.40^2 + \cdots + 0.20^2 + \cdots + (-0.34)^2) \\ &= 12(1.4004) \\ &= 16.80 \end{aligned}$$

Size of effect

This is expressed absolutely as SS and otherwise as a proportion of the appropriate total SS. If the two component factors of the interaction and the factor whose level is specified are all between-subjects then the between-subjects total SS is appropriate. Otherwise the within-subjects total SS is appropriate.

For example, the simple interaction of **digit × speed** for the dyslexic children only accounts for 16.80/694.24 = 2.42% of the within-subjects variation.

Test of significance

The F value is formed by dividing the MS of the simple interaction by the appropriate MS_{error}. The MS_{error} is $MS_{within\text{-}groups}$ for the BBB design. For other designs it is difficult to identify. The reader is referred to Winer *et al.* (1991, section 7.4).

13.5.6 Contrasts and comparisons in three-factor designs

Introduction

Contrasts and comparisons may be tested among the levels of the factors in the three-factor designs in the same way as in the two-factor designs, as discussed in section 12.7.

The only area of possible difficulty is the identification of the appropriate error MS for the test of significance. The simple rule is that the same MS_{error} is used for testing a contrast among the levels of a factor as is used for testing the main effect of the factor. This is tabulated in section 13.5.7.

As mentioned in Chapter 8 and section 12.7, no *a posteriori* comparisons are available for within-subjects factors.

Example of a contrast in the BWW design

Consider again the example of the BWW experiment on errors in recall for digits in three-digit numbers. Suppose there was a requirement to test the *a priori* contrast of *first* and *third* digit positions against the *second* digit position (primacy and recency memory effect versus the other). The linear contrast function would have the value

$$(-1)(0.85)+(+2)(1.65)+(-1)(2.83)=-0.38$$

and SS the value

$$\frac{(96)(0.38)^2}{1^2+2^2+1^2}=2.3104$$

The multiplier n has the value 96. This is because each mean, such as 0.85, is based on 96 units of measurement, i.e. the mean number of errors (which serves as the dependent variable).

The appropriate error MS is $MS_{subjects \times digit}$ since this is a within-subjects factor and so tested using $MS_{\text{subjects} \times \text{factor}}$. This is the error MS used in the test of **digit** main effect.

Hence $F = 2.310/0.91 = 2.538$ with 1 and 44 degrees of freedom. Since F is less than F_c the decision is made not to reject H_0.

13.5.7 Summary of error terms in three-factor designs

Using an obvious notation the error terms used to test the main effects and interactions are:

	Error terms	*Sources tested*
BBB	within-groups	all factors, interactions and simple effects
BBW	subjects	B1, B2, B1 × B2
	subjects × W	W, W × B1, W × B2, W × B1 × B2
BWW	subjects	B
	subjects × W1	W1, B × W1
	subjects × W2	W2, B × W2
	subjects × W1 × W2	W1 × W2, B × W1 × W2
WWW	subjects × W1	W1
	subjects × W2	W2
	subjects × W3	W3
	subjects × W1 × W2	W1 × W2
	subjects × W1 × W3	W1 × W3
	subjects × W2 × W3	W2 × W3
	subjects × W1 × W2 × W3	W1 × W2 × W3

13.6 EXERCISES

13.1 A study of eye movements in dyslexic and normal children was designed as a factorial ANOVA with three factors all at two levels.

Factor **A** describes the type of child, level 1 for dyslexic, level 2 for normal. There were 15 children of each type.

Factors **B** and **C** describe the conditions under which measurements were made. Each child was tested at all levels of both factors **B** and **C**.

The dependent variable is the number of corrections in the movement of the eyes while they follow a moving spot.

The two levels of factor **B** describe the speed of movement of the spot, fast or slow.

The two levels of factor **C** describe the direction of travel of the spot, left-to-right or right-to-left.

The ANOVA summary table is set out below:

Source	*SS*	*df*
A	2.13	1
Subjects (s)	354.6	28
B	55.4	1
C	4.03	1
BC	7.75	1
AB	0.102	1
AC	0.833	1
ABC	0.169	1
sB	264.0	28
sC	13.2	28
sBC	7.07	28
Total	709.2	119

The tables of marginal means are set out below:

	Fast	**Slow**	Means
Dyslexic	4.175	2.758	3.467
Normal	3.850	2.550	3.200
Means	4.013	2.654	

	L to R	**R to L**	Means
Dyslexic	3.367	3.567	3.467
Normal	2.933	3.467	3.200
Means	3.150	3.517	

	L to R	**R to L**	Means
Fast	3.575	4.450	4.013
Slow	2.725	2.583	2.654
Means	3.150	3.517	

(a) Carry out tests of significance of all factors and interactions.
(b) Summarize the findings in simple English.
(c) Represent on a bar chart the effect of **speed** for travel of the spot in direction *R to L*.

13.2 An experiment was carried out to assess the effects of sex, certain verbal and visual tasks and delay on recognition of odours presented during completion of the tasks.

Male and female subjects were randomly allocated to one or other of the three conditions:

1. Verbal task
2. Visual task
3. No task

During completion of one of the tasks or the no-task condition, 12 odours were presented at 2-minute intervals. The order of presentation was randomized for each subject. Subjects were tested on their recognition of the 12 odours out of a collection of 30 odours. A score in the range 0 to 12 was obtained which was the number of odours correctly identified (adjusted to take account of mis-identification). Subjects were tested one hour and one week after the initial presentation of the odours.

There were 4 male and 4 female subjects taking part in each of the 3 conditions (24 subjects in total). The mean numbers of odours recalled at each combination of the levels of the independent variables are set out in Table A.

Table A

	Condition	**Hour**	**Week**	
	1	3.50	7.00	
Male	2	5.00	5.75	mean 6.042
	3	8.25	6.75	
	1	6.75	4.50	
Female	2	8.50	6.75	mean 7.500
	3	10.25	8.25	
	Mean	7.042	6.500	

Table B

Source	*df*	*SS*	*MS*
Sex	1	25.5	25.5
Condition	2	70.8	35.4
Sex × Condition	2	7.5	7.5
Occasion	1	3.5	3.5
Occasion × Sex	1	25.5	25.5
Occasion × Condition	2	11.3	5.65
Occasion × Sex × Condition	2	14.0	7.0
Subjects	18	173.1	9.62
Subjects × Occasion	18	35.1	1.95

Within the general aims of the research it was required to examine the following propositions:

(i) The conditions differ in their influence on the number of odours recalled.

(ii) The extent to which recall diminishes over the first week is different for males and females.
(iii) The change in recall over the first week in condition 1 (verbal task) depends on **sex**.
(iv) In the control condition there is a reduction in recall over the first week.

By reference to Tables A and B for each of the propositions (i) to (iv):

(a) Display the appropriate means in a graph.
(b) Obtain the relevant *SS* and express it as a percentage of the appropriate total *SS*.

For each of the propositions (i) and (ii):

(c) Perform the appropriate F-test of significance and report the results.

Appendix A: Hints on use of computer programs

This note is based on the author's experience as a working statistician rather than on an exhaustive survey of available programs. Accordingly, many programs favoured by readers will not be mentioned. This does not imply anything about their suitability for analysing the designs dealt with in this book.

A.1 SAS and SPSS

These are available on a range of platforms including mainframe and personal computers. They are capable of carrying out all analyses dealt with in this book. Both require an above-average investment of effort by the new user, but that is the price to be paid for the comprehensive coverage and wealth of facilities provided. Many of the benefits of these systems of programs go to the user with a large amount of data who requires many analyses.

There is a degree of inconvenience associated with their use for analysis of simple effects and of within-subjects factors.

A.2 STATVIEW

This is designed for the Apple Macintosh. It is extremely easy to use and covers most of the more basic designs dealt with in the book. The limitations are that it cannot accept more than one within-subjects factor in a design, unbalanced designs or analysis of covariance.

A.3 SUPERANOVA

This is designed to be compatible with Statview. It is designed for the Apple Macintosh and is easy to use. It analyses all the designs dealt with in the book with the small exception that it does not provide a stepwise version of multiple regression.

SuperANOVA is particularly useful for its handling of simple effects, unbalanced designs and for plotting bar charts and interaction diagrams.

A.4 MINITAB VERSION 8

This is available on most platforms including mainframes, IBM PCs and Apple Macintoshes. It can be used to analyse most of the designs dealt with in the text. An important exception is that it deals with designs with within-subjects factors in a different way. 'Subjects' has to be introduced as a 'random' factor 'nested' under all between-subjects factors and 'crossed' with all within-subjects factors. Otherwise it is easy to use and flexible. It is available at a reduced price in a student edition through bookshops.

A.5 BMDP

This is available for mainframe and IBM PC platforms. It can be used to analyse all designs dealt with in the text. The BMD suite of programs is particularly valuable for the user of designs which include within-subjects factors.

A.6 SYSTAT

This is designed for the Apple Macintosh. It is quite easy to use and can be used for analysis of all the designs dealt with in the text.

A.7 STATISTICA (CSS)

This is designed for the IBM PC. It can be used for analysis of all designs dealt with in the text.

Appendix B: Additional exercises for Chapters 5–13

CHAPTER 5

B5.1 Consider a single-factor repeated measures design in which there are three levels of the **treatment** factor and five **subjects**. The overall mean score on the dependent variable is 26.3 and the effects present in the design expressed as deviations from the overall mean are as follows:

Treatment factor: (+5.3, +2.9, −8.2)

Subjects:

S_1	−10.3
S_2	+2.5
S_3	+13.4
S_4	+0.2
S_5	−5.8

Interaction:

+1.5	−0.9	−0.6
+0.3	+1.7	−2.0
−2.4	−2.0	+4.4
−1.6	−0.2	+1.8
+2.2	+1.4	−3.6

(a) Obtain the values of the scores obtained by **subject** number 1 under each of the three **treatment** levels. Sketch a bar chart to display these values.
Obtain the **treatment** effect as experienced by the first **subject** as a set of deviations from the overall mean.
(b) Identify the **subject** most affected and the **subject** least affected by the **treatment** factor.
(c) Calculate the *SS*s for the effect of the **treatment** factor and for the **subject** × **treatment** interaction. Complete the *F*-test for the **treatment** factor.
(d) Carry out an *a priori* test for trend for the **treatment** factor.

B5.2 Choice reaction times in tenths of a second were obtained for each of four subjects in three different test **conditions** (**C1**, **C2** and **C3**).

Each subject was tested under all three **conditions** in random order. The reaction times were:

Subject	**C1**	**C2**	**C3**
1	12	8	8
2	7	5	6
3	15	12	5
4	10	8	6

Overall mean = 8.5.

(a) Express the **subjects** and **conditions** effects as sets of deviations.
(b) Remove the **subjects** and **conditions** effects and the overall mean from the reaction times so as to expose the **subject** × **conditions** interaction as an array of deviations.
(c) Comment on the extent to which individual subjects vary one from another in their patterns of response to the conditions.
(d) Complete the ANOVA summary table and F-test by taking advantage of the results of parts (a) and (b) of this question.

CHAPTER 6

B6.1 This exercise is based on data from Postman and Keppel (1977).

Mode of **presentation** of stimuli and mode of **response** were investigated for their effects on recall of pairing of items from a list. Two modes were used, picture (**P**) and word (**W**).

Eighteen subjects were assigned to each of four versions of the list. The first group of 18 subjects learned the list in picture mode and was tested one week later, again in picture mode. This is the **P–P** condition. The other three groups underwent conditions **P–W**, **W–P** and **W–W**.

The mean numbers of items recalled correctly was:

P–P	9.94
P–W	19.06
W–P	12.39
W–W	15.06

Rearranged so as to display the effects of **presentation** and **response** through the appropriate marginal means, the means appear thus:

		Response		
		P	**W**	
Presentation	**P**	9.94	19.06	14.50
	W	12.39	15.06	13.725
		11.165	17.06	14.113

Rearranged again and expressed as deviations the means are displayed as an additive model thus:

$$\begin{Bmatrix}\text{recall}\\ \text{score}\end{Bmatrix} = 14.113 + \begin{Bmatrix}+0.3875\\ -0.3875\end{Bmatrix} + \begin{Bmatrix}-2.9475\\ +2.9475\end{Bmatrix} + \text{interaction}$$

Presentation {**P**} {**W**} **Response** {**P**} {**W**}

(a) Express **interaction** in the form of deviations appropriate for inclusion in the model.
(b) Given that the *SS* within-groups is 4095, calculate all other *SS*s, complete the ANOVA table and report the tests of significance of **Presentation**, **Response** and interaction.
(c) Summarize the results of the experiment in simple English and with appropriate sketch graphs.
(d) Does the mode of response make a significant difference to the recall when the presentation is in picture mode?

B6.2 An experiment was carried out to study the effects on body weight of four diets alone or combined with group discussion therapy.

Forty men who wished to lose weight were randomly allocated to groups of five. Each of the eight groups was randomly allocated to a unique combination of one of the diets and either discussion therapy or no discussion therapy. The men were weighed on starting the diet and again after following the diet for three months.

Weight loss was calculated by subtracting the final weight from the original weight of each man and was used as the dependent variable in the analysis of the results.

The mean weight lost in pounds by each group of men is set out in Table B.1.

Table B.1

Rows: **therapy**		Columns: **diet**				Means
		1	**2**	**3**	**4**	
Therapy	1	14.8	21.2	19.6	19.2	18.70
No therapy	2	9.4	13.4	14.8	29.8	16.85
Means		12.1	17.3	17.2	24.5	17.775

(a) The following propositions were to be examined:

(i) Group discussion **therapy** helps men to lose weight.
(ii) The effect of **diet** on weight loss depends on whether or not the diet is accompanied by attendance at group **therapy**.

For each of these propositions:

- Sketch appropriate graphs to illustrate the effect.

- Choose the appropriate *SS* s from the ANOVA summary in Table B.2 and use them to carry out the corresponding tests of significance.
- Report the results in simple English.

Table B.2 Analysis of variance for weight loss

Source	*SS*
Therapy	34.22
Diet	779.87
Therapy × Diet	529.28
Within-groups	187.60
Total	1530.97

(b) (i) Identify the largest of the simple effects of **therapy** and the smallest of the simple effects of **diet**.

(ii) Test for significance the effect of **diet** in the absence of group discussion therapy.

(c) Discuss the implications of a significant interaction for the interpretation of main effects in designs such as the one used here.

B6.3 An experiment was carried out to investigate subjects' perceptions of a fictitious member of the public (the target person) described as either **young** or **old** and as having reacted in an **aggressive**, **passive** or **mixed** way to an unwanted approach in the street.

The experimental subjects had to select traits from a list to describe the target person. The dependent variable is the percentage of the selected traits that are masculine (as rated previously by a panel).

A sample of 30 subjects was allocated at random, five to each of the six different combinations of descriptions.

The results, expressed as an additive model are:

$$\text{mean percentage masculine traits} = 53.4 + \left\{\begin{matrix} +17.4 \\ -8.1 \\ -9.3 \end{matrix}\right\} + \left\{\begin{matrix} -18.8 \\ +18.8 \end{matrix}\right\} + \left\{\begin{matrix} +1.3 & -1.3 \\ +6.0 & -6.0 \\ -7.3 & +7.3 \end{matrix}\right\}$$

response {**aggressive**, **passive**, **mixed**}; **age** {**young**, **old**}; **interaction**

(a) Obtain the appropriate means and express graphically the main effect of **response** and the interaction of **response** with **age**.

(b) Given that the *SS* within groups is 19 264.8, calculate further *SS* s and complete the summary table. Carry out the test of significance of **response** main effect.

(c) Test for significance the smallest simple effect of **age**.

(d) Summarize the main findings of the study in terms likely to be used by the experimenter.

B6.4 The result of a two-factor experiment was expressed in terms of the deviations of the main effects and interactions as:

$$\text{score} = 6.77 \quad + \quad \begin{Bmatrix} -1.6 \\ +1.6 \end{Bmatrix} \quad + \quad \begin{Bmatrix} +0.98 \\ -1.12 \\ +0.13 \end{Bmatrix} \quad + \quad \begin{Bmatrix} +1.95 & +0.25 & -2.2 \\ -1.95 & -0.25 & +2.2 \end{Bmatrix}$$

Factor **A** Factor **B** Interaction **A** × **B**

(a) Calculate the score expected for a subject responding under the influence of the first level of factor **A** and the third level of factor **B**.
(b) Given that the experiment was an independent groups design with five subjects per group (hence 30 subjects in all) and on the basis of a simple calculation, decide which of factors **A** and **B** and their interaction makes the greatest contribution and which the smallest contribution to explaining the total variation in scores.
(c) If the two simple effects of factor **B** expressed as deviations are (2.93, −0.87, −2.07) and (−0.97, −1.37, 2.33), respectively, show that the *SS* interaction can be obtained as the variation of the simple effects from their mean.
(d) What well-known concept can be viewed as the average of the simple effects? Demonstrate this numerically.
(e) Explain what is meant by the statement that the factors **A** and **B** and their interaction make *independent* contributions to explaining the variation in scores of the dependent variable.

CHAPTER 7

B7.1 An experiment was carried out on the effect of practice on picture recognition. Twenty children were randomly allocated to two groups of ten. One group received practice, the other did not. Both groups were tested on recall of 40 pictures.

The mean numbers of pictures recalled in the two groups and their corresponding mean IQ scores were:

Condition group	*Mean* **recall**	*Mean* **IQ**
Practice	19.90	101.4
No-practice	15.80	96.0

IQ was to be used as a covariate. The *SS*s for the **conditions** factor and the covariate are set out in the Venn diagram (see Chapter 10) of Fig. B.1.

The gradient of the common regression line is 0.388.

(a) Construct ANOVA summary tables for both the adjusted and unadjusted versions and complete both *F*-tests for the appropriate directional hypothesis.
(b) Obtain the adjusted values of the groups' means corresponding to an **IQ** of 100.

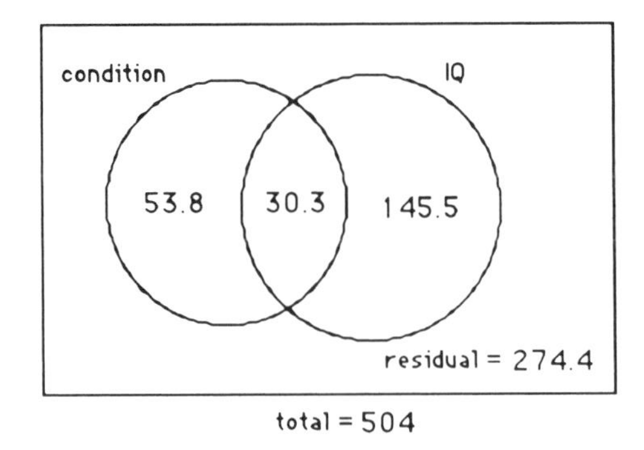

Fig. B.1

(c) Obtain the correlation of **IQ** with the picture recognition score.
(d) Summarize the results of the experiment and state your view on the usefulness of the covariate.

CHAPTER 8

B8.1 Sixty family triads (mother, father, child) took part in a study of the relationship of **family interaction patterns** with **family adjustment** and the **status of the father** (stepfather versus biological).

The families were randomly sampled from populations defined as follows:

(a) biological father and well-adjusted family
(b) biological father and not well-adjusted family
(c) stepfather and well-adjusted family
(d) stepfather and not well-adjusted family

There were 15 families in each group.

Measures were obtained on three variables which related to family interaction patterns:

Variable 1: child's **positive involvement**
Variable 2: **marital relations** quality scale
Variable 3: **family agreement**

All scales were constructed so that a higher score corresponds to more positive involvement, better marital adjustment or to more family agreement.

The researchers made certain predictions about the outcome of the study, which are as follows:

(i) Well-adjusted families will have higher scores on the **marital relations** scale than not well-adjusted families.
(ii) Children in stepfamilies will have a lower **positive involvement** with their father than children in biological families.
(iii) Children in well-adjusted stepfamilies will have greater **positive involvement** with their fathers than children in not well-adjusted stepfamilies.

(iv) **Family agreement** will be greater in well-adjusted than in not well-adjusted families.

(a) By reference to the cell means in Table B.3 display graphically the effects appropriate for each proposition (i) to (iv).
(b) By reference to Table B.4 or otherwise, carry out tests of significance of the propositions (i) to (iii), paying particular attention to techniques required for directional F-tests and for simple effects.
(c) Explain why the interaction present among the effects for **family agreement** indicates a difficulty in interpretation of the results in respect of proposition (iv). Carry out a test of the largest simple effect involving this variable.

Table B.3 Mean scores on three scales for four groups of families

Variable	Biological father		Step-father	
	Well-adjusted	Not well-adjusted	Well-adjusted	Not well-adjusted
Child's **positive involvement** with father	13.31	7.93	8.44	3.02
Marital relations scale	114.31	80.67	124.81	102.38
Family agreement scale	23.50	23.07	29.00	20.87

Table B.4 Mean-squares and F-values from ANOVAs for each of three dependent variables

Variable	*MS for error*	*F for* **family adjustment**	*F for* **status of father**	*F for* interaction
Child's **positive involvement** with father	15.05	29.41	24.09	0.01
Marital relations scale	281.6	43.52	14.28	1.75
Family agreement scale	19.45	11.72	3.86	4.42

B8.2 This exercise is based on data from Armitage (1987).

An experiment was carried out to compare the effects of three drug treatments and a control condition on the clotting times of blood plasma. Eight subjects were tested, each under all four treatments.

The mean clotting times (in minutes) were:

Condition	*Drug 1*	*Drug 2*	*Drug 3*	*Control*
Mean time	9.30	9.71	9.94	11.02

(a) Express the **conditions** effects as deviations and hence calculate the *SS* for **conditions**.
(b) Given that the *SS* for subjects = 78.99 and *SS* for interaction = 13.77, complete the overall ANOVA table and test of significance of **conditions**.
(c) Carry out the *a priori* test of the hypotheses:

H_0: conditions do not differ
H_1: drug conditions (considered together) lead to lower clotting times than the control condition.

(d) Give your interpretation in simple English of the **subjects** × **conditions** interaction.

CHAPTER 9

B9.1 Compare the usefulness of the concepts of power and sensitivity to the designers of experiments.

B9.2

(a) In a randomized independent groups design with three groups, the 'true' effect of a factor on the values of the dependent variable is (5, 12, −17). The result of an experiment designed to evaluate the effect is the following set of means:

$G_1 = 50.92$
$G_2 = 55.12$
$G_3 = 25.48$

If there are $n = 5$ subjects per group, the within groups *SS* is 4101 and the significance level for the test is 0.05, calculate the power of the test.

(b) (i) For the experiment in part (a) obtain the sensitivity and the efficiency. Assume the cost, in terms of subject time, is 0.25 hours per subject and, in terms of researcher time, is 0.5 hours per subject plus a once-and-for-all 6 hours set-up time.
(ii) Identify the influences on the power of a single factor independent groups ANOVA with reference to the example in part (a).
Explain how the power may be improved without losing generalizability or economy.

B9.3 A pilot study is to be carried out into the effect of certain therapeutic activities on mood in an institution-resident elderly population. Individuals are to experience one of the following three conditions:

Condition 1: Listen to unfamiliar songs
Condition 2: Listen to familiar songs
Condition 3: Sing along to familiar songs

The dependent variable **mood** is measured on a scale from 0 to 25 using a standard interview.

The comparison of conditions 2 and 3, which use the same songs, is to investigate the active versus passive aspect.

The comparison of conditions 1 and 2 is to investigate the familiar versus unfamiliar aspect.

The presence or absence of dementia was thought likely to influence **mood**.

Accordingly, the design of the pilot study was as follows. By random selection, three groups of individuals were assembled each consisting of 4 **dementia** and 4 **non-dementia** residents. Each group was assigned to one of the conditions thus:

	Condition		
	1	**2**	**3**
Dementia	$n=4$	$n=4$	$n=4$
Non-dementia	$n=4$	$n=4$	$n=4$

The time costs of the pilot study were as follows:

Running the music sessions 3 @ 2 h	= 6 hours
Mood testing, each individual 24 @ $\frac{1}{2}$ h	= 12 hours
Dementia testing 24 @ $\frac{1}{4}$ h	= 6 hours
Fixed set-up costs	= 10 hours
Total	34 hours

The results of the two-factor analysis of variance in terms of sums of squares (SSs) are displayed on the Venn diagram in Fig. B.2.

(a) Set out the complete analysis of variance table and report the test of significance of **conditions**.

(b) One of the main reasons for the pilot study was to explore the usefulness of the **dementia** factor. Apply appropriate methods to compare the pilot study as run with a hypothetical alternative version in which consideration of dementia is omitted. Comparison of efficiency and sensitivity should be among the approaches you adopt.

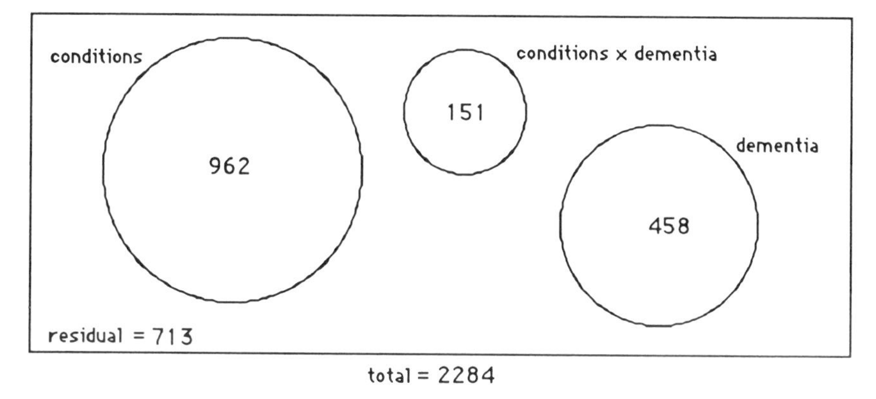

Fig. B.2

(c) Does a version of the design which ignores **dementia** and retains 24 subjects in total have enough power for the test of significance of a **conditions** effect of size (−7, +4, +3)?

B9.4 An experiment was set up to compare the levels of anxiety generated by three alternative methods of extracting wisdom teeth.

Thirty patients needing extractions were randomly allocated, each to one of the methods, so that ten patients were treated by each method.

The Spielberger state anxiety score was obtained 15 minutes after the extraction and served as the dependent variable for the analyses of the results.

The mean **anxiety state** scores for the three methods were:

Method	*Mean*
1	33.48
2	32.58
3	33.85

The *SS*s obtained from an analysis of variance of the state anxiety scores were:

Source of variation	*SS*
Between methods	8.533
Within groups	54.657

Two potentially useful covariates were available, the ages of the patients (in years) and trait anxiety scores obtained before the patients were told of the need for an extraction.

(a) (i) Obtain the relative efficiencies of the versions of the experiment with each of the covariates as compared to the basic version of the experiment without covariates.

The necessary information about *SS*s and costs follows. The costs, in terms of time spent by the researcher were:

Trait anxiety measurement	20 min	per patient
State anxiety measurement	20 min	per patient
Age in years	5 min	per patient
Set up and analysis time	6 hours (once only)	

The *SS*s were:

Source of variation	*SS adjusted for* **age**	*SS adjusted for* **trait**
Between methods	8.566	19.98
Within groups	52.757	36.40

(ii) Report and justify your conclusions about the usefulness of the covariates.

(b) The model which includes trait anxiety score as covariate has the form:

$$\text{state score} = 33.303 + \begin{Bmatrix} -0.089 \\ -1.076 \\ +1.165 \end{Bmatrix} + (0.5424)\ (\text{trait} - 31.40)$$

The overall mean state score is 33.303 and the overall mean trait score is 31.40.

Interpret the three terms in the model in language that would be clear to a dentist. Illustrate your interpretation by using the model to calculate the predicted state anxiety score for a fictitious patient.

(c) Carry out and report tests of significance for all adjusted and unadjusted models. Discuss the usefulness of the covariates in view of the results of these significance tests.

B9.5 Sixty dysfunctional families took part in an experiment which aimed to compare different approaches to family therapy. The families were randomly allocated to four groups of 15.

Each group of families was randomly allocated to one therapy arrangement. Two independent factors were under investigation:

Approach	(two levels)	behaviour-oriented
		feelings-oriented
Location	(two levels)	the family home
		the clinic

The dependent variable was the change (after − before) in **family agreement score** over the period of the experiment.

The design layout of the experiment displaying the numbers of families and mean change scores follows:

		Location	
		family home	**clinic**
Approach	**behaviour-oriented**	n = 15 23.50	n = 15 23.07
	feelings-oriented	n = 15 29.00	n = 15 20.87

An extraneous variable thought to be useful for increasing the power was the total number of years post-15 education of the parents. This variable is referred to as **education**.

The Venn diagram in Fig. B.3 displays the *SS*s due to the two experimental independent variables and the covariate.

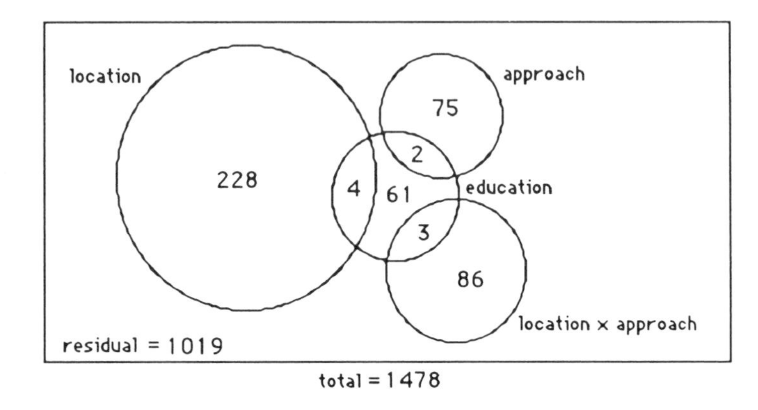

Fig. B.3

The cost of the experiment can be assumed made up thus:

50 hours to set up the experiment
4 hours for each family. (The time in therapy would have happened anyway so was not part of the cost of the experiment.)
0.25 hours per family to obtain the information for the **education** variable.
6 hours additional work involved in the ANCOVA analysis if **education** used, additional to the work of carrying out the ordinary ANOVA if **education** was not used.

(a) Apply the following three methods for comparing versions of the experiment with and without use of the extraneous variable **education**:
   (i) Rule of thumb assessment of strength of relationship of extraneous variable to dependent variable.
   (ii) Comparison of sensitivities of both versions.
   (iii) Comparison of efficiencies.

   Give your assessment as to whether it is worth using the **education** variable.

(b) (i) Calculate the powers for the tests of significance of each of the main effects in the experiment. Take the means in the experiment as estimates of the sizes of the effects of the factors, i.e. take the deviations to be:

   approach $(-0.825, \quad +0.825)$
   location $(-2.140, \quad +2.140)$

   (ii) Comment on the adequacy of the power values obtained in (i).

(c) Discuss the reasons for the experimenter's making use of random allocation in the above experiment rather than working with families that just happened to be in one or other arrangement of therapy.

B9.6 An occupational psychologist carried out an experiment on the effect of different organizational arrangements on the profitability of a firm's retail shops. Fifteen of the company's shops, chosen at random, were allocated, at random, each to one of three different organizational arrangements. In addition to the mean profit (in £1000s) for the three-month period of the experiment, a covariate measure was obtained which was the number of square feet (in 1000s sq. ft.) of retail floor of each shop.

The results are set out in Tables B.5 and B.6.

Table B.5 Scores and mean scores of the dependent variable and covariate

Organization 1			**Organization 2**			**Organization 3**		
Shop No.	**Retail space**	**Profit**	**Shop No.**	**Retail space**	**Profit**	**Shop No.**	**Retail space**	**Profit**
1	26.8	22.2	6	40.7	22.5	11	13.6	34.8
2	38.1	25.2	7	5.2	15.9	12	41.2	39.2
3	44.5	26.8	8	24.4	19.6	13	17.9	33.4
4	25.6	33.3	9	42.6	35.5	14	18.6	38.8
5	12.0	15.7	10	47.2	40.4	15	9.1	20.0
Mean	29.4	24.6		32.0	26.8		20.1	33.2

Table B.6 ANCOVA summary table

Source	*df*	*SS*	df_{adj}	SS_{adj}
Region	2	200.452	2	421.502
Residual	12	859.832	11	426.913
Total	14	1060.284	13	848.415

(a) Given that the regression gradient is 0.421, calculate the mean profits of each of the three groups of shops adjusted to what they would be if all shops had exactly 27.17 thousand sq. ft. of sales space.
(b) Interpret the results of the experiment so as to reach a decision about the relative merits of the three organizational methods and so as to clarify the role of the covariate.
(c) If collating together the profit figures for one shop takes on average four staff hours, and obtaining the area of sales space takes six staff hours, investigate the relative efficiency of the ANCOVA as compared to the equivalent simple one-factor ANOVA.

B9.7 An experiment was carried out to study the effects of four **drugs** on reaction time to a series of standardized tasks. All subjects had been given extensive training on those tasks prior to the experiment. The 16 subjects used in the experiment were a random sample from the appropriate population, eight male and eight female. **Sex** was taken account of in the design and analysis, solely as a blocking factor in order to increase the power of the test of the **drugs** treatment factor.

All subjects' reaction times were measured under the influence of each of the four drugs. The order of treatment was randomized for each subject.

The ANOVA summary table was:

Sources within subjects	*SS*	*Sources between subjects*	*SS*
drugs	591.6	**sex**	54.0
drugs × **sex**	1544.2	residual (**subjects**)	680.8
residual (**subjects** × **drugs**)	1128.8		

(a) Sketch the Venn diagram showing the apportionment of *SS* appropriate to the analysis.
(b) Set out the ANOVA summary table and corresponding Venn diagram that would be expected in an alternative version of the experiment in which the blocking factor, **sex**, was ignored.
(c) Compare the significance of the effect of the **drugs** factor in the two versions of the experiment.
(d) Calculate the sensitivity of the test of **drugs** for both versions of the experiment. Assuming the costs of the two versions of the experiment are identical, calculate their relative efficiency.
(e) What sample size would be needed in the unblocked version of the experiment to give the same sensitivity as that obtained in the randomized block version?
(f) Identify the implications of carrying out this experiment using only female subjects.

CHAPTER 10

B10.1 A survey was carried out on a sample of nursing undergraduates.

(a) Part of the survey involved relating science A level attainment to students' attitudes to their physiology courses. Attitude was measured so that a large score indicated a favourable attitude. The attitude scores of the individual students grouped according to **sex** and A level result were:

Sex	Science A level passed	Attitude scores								*n*
Female	Yes	15	18							2
	No	13	10	11	14	9	17	15	10	8
Male	Yes	15	9	12	12	10	14	10		7
	No	11	16	12						3

(i) Obtain the mean **attitude** scores for the science A level and no science A level groups both unadjusted and adjusted for the effect of **sex**.
(ii) Interpret the result.

(b) One further item of information available for each of the 20 students was percentage of physiology classes attended throughout the course. This information is presented on a scatter plot against the attitude to physiology score (Fig. B.4) separately for the science and no science students. The students who have passed science A level are plotted with a dot in a square, the other students with a diamond. The mean percentage of classes attended were 58.5% and 61.5% in the science and no-science groups respectively. The appropriate parallel regression lines are drawn on the scatter plot.

(i) Use these lines to estimate the adjusted mean attitude scores in the two groups (adjusted, that is, to the value 60% attendance).
(ii) Evaluate the effect of the adjustment.

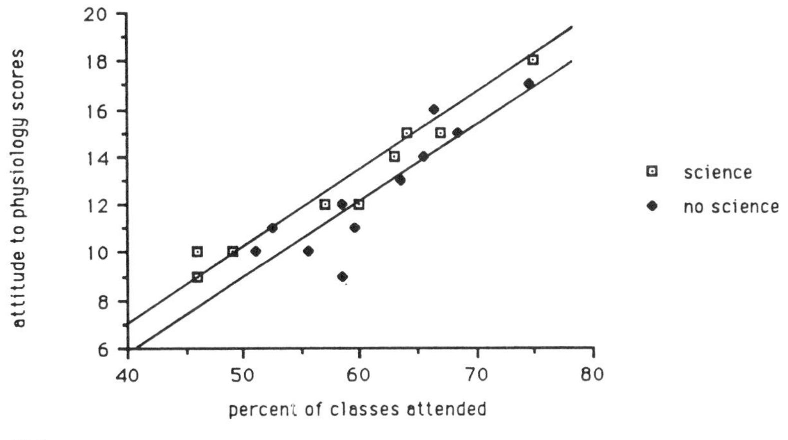

Fig. B.4

B10.2 As part of the World Fertility Survey in Fiji it was required to compare the mean number of children per woman in rural and urban **areas**. Data for 16 women are set out below, classified by number of **years** since marriage as well as by type of **area**. The number of children and, in brackets, the number of **years** of schooling, are given for each woman.

Years *since marriage*	*Urban*				*Rural*		
<5	2(4)	0(6)	1(4)		4(3)	0(4)	
5–9	7(3)	2(5)			5(2)	8(2)	6(2)
10–14	3(3)	6(3)	0(7)	6(2)	9(2)	4(5)	
Means		3.00(4.11)				5.14(2.86)	

(a) Obtain the mean numbers of children to urban and rural women adjusted for number of years since marriage (i.e. adjusted to what they could be expected to be if the urban and rural groups had the same numbers in each category of years since marriage).

Comment on the effect of the adjustment you have carried out.

(b) The number of children to each mother was used as the dependent variable in an analysis of variance. The total *SS* and amounts of *SS* explained by **years** since marriage and **area** of residence are indicated on the Venn diagram in Fig. B.5.

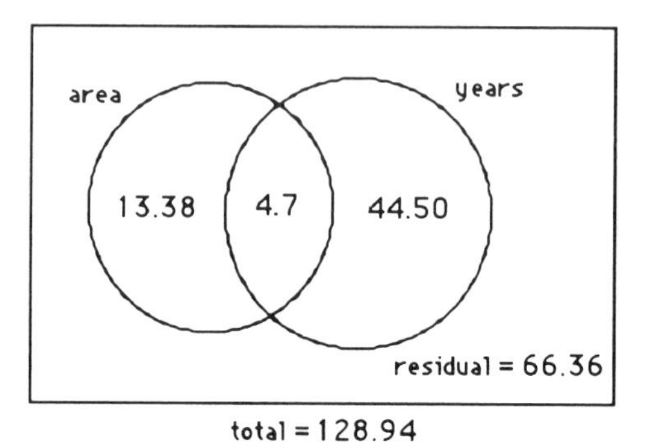

Fig. B.5

Construct appropriate ANOVA summary tables and complete tests of significance for the effect of **area** of residence both adjusted for and not adjusted for the effect of **years** since marriage.

(c) A further analysis of variance was carried out to adjust the numbers of children for the numbers of **years** of schooling of the women. It resulted in the sums of squares indicated on the Venn diagram in Fig. B.6.

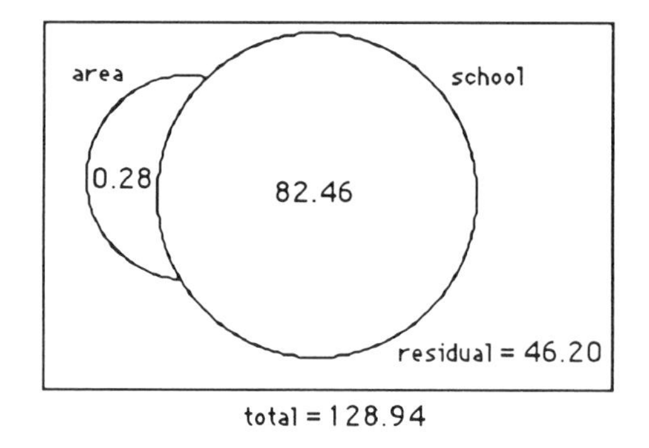

Fig. B.6

Construct an appropriate summary table for the effect of **area** on numbers of children adjusted for the effect of numbers of years of **schooling**. Complete the test of significance.

B10.3 For each of the investigations (a) to (f) below:

(i) Discuss how far the design employed limits the extent to which a causal link can be established of the independent to the dependent variable.

(ii) Identify a possible confounding variable and state its likely effect and the method you would use to adjust for it.
(iii) Suggest design approaches to controlling the effects of the possible confounding variable.

(a) It is required to study the effect of mode of presentation of information about childbirth on anxiety among first-time mothers. Mothers attending three ante-natal clinics will take part in the study. Information will be presented to the mothers as follows:

Clinic 1: booklet
Clinic 2: talk by midwife
Clinic 3: booklet and talk by midwife

The dependent variable is the score on an anxiety questionnaire.
(b) A study is to be carried out of the effect of their earliest experiences of drinking alcohol on students' current attitudes to alcohol consumption on campus. Enjoyment of the first experience was recalled by the students and rated on a scale 0 to 10. This is to be related to current attitude (favourable/unfavourable) on a scale from 1 to 7.
(c) A survey is carried out at a local mother and baby clinic in order to compare growth rates of babies according to whether or not the mothers themselves follow a vegetarian diet. Assume growth rates are measured in grams per month.
(d) An occupational psychologist was employed to study methods of organizing production in garment factories. Two main methods were to be compared:
   (i) the whole garment is completed by one worker.
   (ii) hemming, elastication and finishing are carried out by different workers who specialize.

Twenty factories of the first type and 30 of the second were to be compared with regard to the proportion of factories that had increased their output over the last 12 months.
(e) A survey was to be carried out among women living in an inner city area to relate their ages to their fears about personal safety when going out at night. The women's level of fear was to be measured on a continuous scale from 0 to 19.
(f) All 15 health centres in Bloomsbury and Islington Health District were included in an investigation of the effect of certain optional facilities on customers' self-reported level of satisfaction with the service received. The facilities were:

   (i) Presence of nursery with nurse
   (ii) Presence of drinks machine in waiting area
   (iii) Presence of coin-box telephone

Satisfaction was measured on a continuous scale from 0 to 10.

B10.4 Identify the validity problem of quasi-experiments and explain how it is ameliorated.

CHAPTER 11

B11.1 A multiple regression analysis was used to investigate the relationship between girls' interest in sport and their parents' interest in sport. This was part of a wider study into the development of sexual stereotyping. The girls were first-born children and aged 14 years. Identical questionnaires were used for the parents and daughters and gave a score out of 50. An additional measure was the girl's general athletic ability rated, out of 20, by her sports teacher.

Table B.7

	Correlations		
	Y	X_1	X_2
X_1	0.607		
X_2	0.465	0.772	
X_3	0.793	0.424	0.382

Table B.8

Number of variables in model	*R-squared*	*Variables in model*
1	0.216	X_2
1	0.368	X_1
1	0.629	X_3
2	0.368	X_1 X_2
2	0.660	X_2 X_3
2	0.718	X_1 X_3
3	0.722	X_1 X_2 X_3

(a) With reference to the correlations in Table B.7 and the complete set of multiple R-squareds in Table B.8 analyse the contribution to explaining variation in $\mathbf{Y}$ (the daughter's interest in sport) of $\mathbf{X}_1$ (the father's) and $\mathbf{X}_2$ (the mother's interest in sport). Give answers both allowing and not allowing for the explanatory contribution of $\mathbf{X}_3$ (the girl's ability at sport).

Base your analysis on:

(i) Semi-partial squared correlations
(ii) Partial squared correlations

(b) Interpret the coefficients in Table B.9 in a way that contributes to the aims of the research project.

(c) If there were 35 families in the study sample complete the ANOVA table in Table B.10 for the full model and carry out the test of significance.

Table B.9

Variable	*Regression coefficient*
X_1	+0.52
X_2	−0.15
X_3	+1.72

Table B.10

Source	*SS*
X_1, X_2 and X_3	2443
Residual	938
Total	3381

B11.2 An investigation was carried out of the working of a drug used to lower blood pressure (BP) during operations. The extent to which BP was lowered depended on the way in which the drug was administered. Data was obtained in the form of a survey of 53 operations. The variables studied were:

X_1 (recovery time): the length of time in minutes for BP to return to normal after discontinuation of administration of the drug.
X_2 (dose): amount of drug given in milligrams.
X_3 (level): the mean level of BP during administration of the drug.

The correlations among these variables were:

	X_2	X_3
X_1	0.335	−0.108
X_2		0.469

Considering X_1 as the dependent variable, a multiple regression was obtained with an associated multiple R^2 of 0.2018, as follows:

$$X_1 = 23.01 + 23.64X_2 - 0.715X_3$$

(a) Using the separate correlations of X_1 with X_2 and X_3 and the value of multiple R^2, construct the Venn diagram for explained variation in X_1. Label each region by the proportion it represents.
(b) Using the Venn diagram obtained in (a) or otherwise, obtain the correlation of X_1 with X_2, partialling out the effect of X_3. Comment on the effect of the partialling.
(c) Given that the total *SS* of X_1 is 13 791, complete the ANOVA summary table for the multiple regression of X_1 on X_2 and X_3 and test the significance of this full model.
(d) Discuss the relationships among the three variables in terms likely to be used by the researchers who carried out the study. The discussion should be based on interpretations of the Venn diagram, the correlations, the regression equation and the significance test.

B11.3 A multiple regression analysis was carried out as part of a study designed to investigate the relationships between cognitive ability at age five years (**CA5**) and four independent variables. The i.v.s were a measure of family background, experiences up to age three years and from three to five years and verbal ability at three years.

The independent variables can all be regarded as continuous, are labelled X_1 to X_4, and are defined as follows:

X_1 Socio-economic status of family at birth (1 = professional employment group through to 6 = unskilled group).
X_2 Number of developmentally appropriate experiences between ages 1 and 3 years.
X_3 Verbal ability at age 3 years (high score corresponds to high ability level).
X_4 Number of developmentally appropriate experiences between ages 3 and 5 years.

The correlations of the i.v.s with **CA5** were 0.37, 0.25, 0.67 and 0.30, respectively.

The proportions of the total *SS* of the d.v. explained by various combinations of the i.v.s are set out in the following table:

1				0.137
2				0.063
3				0.449
4				0.090
1	2			0.160
1	3			0.456
1	4			0.153
2	3			0.462
2	4			0.110
3	4			0.472
1	2	3		0.476
1	2	4		0.168
1	3	4		0.483
2	3	4		0.459
1	2	3	4	0.490

(a) Identify the order in which variables would be included in the model in a stepwise forward procedure.
(b) Which is the best combination of three i.v.s? Display on a Venn diagram the proportion of the *SS* total explained uniquely by each of them.
(c) What percentage of the total variation in **CA5** was not explainable by the model combining all four i.v.s?

(d) Given that the total *SS* for **CA5** was 591.7 and the analysis was based on a sample of 748 children, set out the ANOVA table for the model based on X_1, X_2 and X_3. Report the result of the test of significance.
(e) Generally evaluate the usefulness of measurements made after the third year.

CHAPTER 12

B12.1 This exercise is based on data from Hazrati (personal communication).
A study was carried out of visual and phonetic coding in deaf and hearing children.
Twenty deaf and twenty hearing children were tested on their recall of words and symbols. There were five lists, each of 12 words or symbols as follows:

List no.	*Description*	*Examples*
1	Control condition nouns	box, shoe, apple
2	Symbols	× >
3	Words not amenable to visual code	from, you
4	Phonetically similar words	true, view
5	Words similar in deaf signing	right, ought

All children were shown and then tested on all five lists. The number of words or symbols recalled was recorded.
The following propositions were to be examined in the light of the results:

(i) Hearing children will perform better than deaf children overall.
(ii) Recall will be better in the control condition than in all other conditions considered together.
(iii) Deaf children will perform better than hearing children in the second condition.
(iv) Deaf children will show a different pattern of response to the conditions than the hearing children.

The mean numbers of words recalled in the various combinations of conditions were:

		Condition				
		Control	2	3	4	5
Group	hearing	8.3	5.0	6.1	7.6	7.8
	deaf	6.8	9.2	6.0	6.5	7.5

The ANOVA summary table was:

Source	*SS*	*df*
Condition (C)	64.32	4
Group (G)	2.88	1
G × C	209.12	4
Subjects	24.77	38
Subjects × C	33.29	152
Total	334.38	199

(a) For each of the propositions (i) to (iv) identify the set of means and sketch the graphical representation which best informs about it.
(b) By calculating *SS*s where necessary, complete the tests of significance for propositions (iii) and (iv).
(c) Display the various sources of variation on a Venn diagram and hence express the size of the variation due to **condition** as a proportion of the appropriate total *SS*.

B12.2 An experiment aimed to investigate the influence of two modes of stimulation (auditory and visual) and two modes of response (manual and speech) on the reaction times of schizophrenic and normal subjects. Each subject was tested under all four conditions.

The results in terms of mean reaction times (seconds) were as follows:

Group	**Mode**			
	AM	VM	AS	VS
Active paranoid	41	41	50	53
Active non-paranoid	64	65	78	70
Withdrawn paranoid	66	62	78	69
Withdrawn non-paranoid	62	59	75	68
Normals	27	32	37	41

(A = auditory, V = visual, M = manual, S = speech)

(a) Display graphically the interaction of **group** with **mode**. After examining the graph consider whether the results could be clarified by any of the following:

(i) A contrast or comparison among the levels of **group**.
(ii) The formation of new factors from among the levels of **mode**.

Show graphically any main effects or interactions you consider to be relevant.

(b) In the original two-factor analysis, the *SS*s were:

Source	*SS*	*Source*	*SS*
Mode	6 797	**Subjects**	62 304
Group	62 347	**Subjects × Mode**	26 614
Mode × Group	1 563		

Given that there were 12 subjects in each group (60 in all), carry out a test of significance of the simple effect of **mode** for the withdrawn paranoid group. Test also the simple effect of **group** for the AM mode.

B12.3 This exercise is based on data from Armitage (1987).

A longitudinal study was carried out of the effect of overcrowding on the prevalence of chest infection in children. Measurements were made of the number of positive readings of pneumoccus at three ages. (Pneumoccus is the preferred indicator of chest infection.)

Families were randomly selected from each of three categories of overcrowding. There were three families in each category. The first child in each family was measured at 1, 2 and 3 years of age. The mean numbers of positive readings were:

		Age		
		1	2	3
Conditions	overcrowded	9.67	25.00	23.67
	crowded	6.33	14.67	12.67
	uncrowded	3.00	5.67	12.33

The *SS*s were:

Source	*SS*
Age (A)	528.2
Conditions (C)	720.9
A × C Interaction	156.9
Subjects	284.4
Subjects × A	286.2

(a) Display the *SS*s on an appropriate Venn diagram and hence express the variation explained by **conditions** as a percentage of the appropriate total.

(b) Complete tests of significance for **age**, **conditions** and **interaction** and summarize the effects of these factors in a way that would be directly usable by policy-makers.

(c) Carry out and report the results of the following tests of comparisons:
   (i) *A priori* trend of increasing prevalence of infection with increasing age.
   (ii) *A posteriori* comparison of the crowded with the uncrowded conditions.

CHAPTER 13

B13.1 In a study of sensitivity to orthographic structures of deaf and hearing subjects, 10 deaf subjects with poor speech, 10 deaf subjects with good speech and 10 hearing subjects were tested on their ability to recognize a target letter as having been present or not present in non-words displayed briefly on a screen. The non-words were presented in random order having been generated according to the two levels of the factor **regularity** (regular versus irregular) and the two levels of the factor Summed Positional Frequency (**SPF**) (high versus low).

All subjects were tested at both levels of both experimental factors.

The dependent variable was the percentage of correct responses.

The sums of squares are set out in the table below:

Source	*df*	*SS*
Group	2	754.92
Subjects	27	1912.71
Regularity	1	974.70
Group × regularity	2	140.33
Subjects × regularity	27	1876.91
SPF	1	484.00
Group × SPF	2	3.62
Subjects × SPF	27	3020.39
Regularity × SPF	1	18.40
Group × regularity × SPF	2	15.00
Subjects × regularity × SPF	27	641.14
Total	119	9842.12

(a) Complete the ANOVA summary table and complete all tests of significance.
(b) If the mean scores for the three levels of factor group were as follows:

hearing = 76.98
deaf (good speech) = 74.66
deaf (poor speech) = 70.90

   (i) Carry out a test for trend in the means for factor group.
   (ii) Carry out an *a posteriori* test of this contrast: (*hearing* versus *combined deaf groups*).

(c) Set out a data layout diagram for this design.

B13.2 An experiment was carried out on the use of the drug diazepam as a tranquillizer for administration to dental patients immediately prior to their treatment. Thirty-four highly anxious patients and the same number of normal patients took part. One of the aims of the research was to study the effect of the drug on patients' motor coordination. This was achieved by requiring all the patients to perform three motor coordination tests, I, II and III, on four occasions:

(a) immediately before
(b) 60 minutes after
(c) 90 minutes after
(d) one week after administration of the drug.

The scores on the tests are the number of tasks completed in a given time. Tables B.11–B.13 set out the mean scores of the two groups of patients under the various combinations of conditions. Table B.14 is the ANOVA summary table.

(a) For each of the propositions listed as (i) to (iv) below,

(1) Display graphically the means appropriate to the investigation.
(2) Carry out the test of significance. (Hint: select the mean square from the ANOVA summary table or, where necessary, calculate the appropriate mean square.)

Table B.11

	Test I	**Test II**	**Test III**	Means
Normal group	61.575	67.075	75.525	68.058
Anxious group	67.275	61.575	79.925	69.592
Means	64.425	64.325	77.725	68.825

Table B.12

	Before **drug**	*After* **drug**			Means
		60 min	*90 min*	*1 week*	
Normal group	71.333	54.467	72.200	74.233	68.058
Anxious group	72.933	55.600	73.233	76.600	69.592
Means	72.133	55.033	72.717	75.417	68.825

Table B.13

	Before **drug**	*After* **drug**			Means
		60 min	*90 min*	*1 week*	
Test I	66.800	52.450	67.150	71.300	64.425
Test II	68.200	50.200	68.200	70.700	64.325
Test III	81.400	62.450	82.800	84.250	77.725
Means	72.133	55.033	72.717	75.417	68.825

Table B.14

Source of variation	*Sum of squares*	*df*	*Mean square*
Main effects			
Test	32 319.0	2	16 159.5
Group	479.0	1	479.0
Occasion	52 989.0	3	17 663.0
Two-way interactions			
Test × group	5 103.0	2	2 551.5
Test × occasion	668.5	6	111.4
Group × occasion	56.5	3	18.8
Three-way interaction			
Test × group × occasion	966.0	6	161.0
Error terms			
Subjects	12 571.0	66	190.5
Test × subjects	7 116.0	132	53.9
Occasion × subjects	3 770.0	198	19.0
Test × occasion × subjects	13 421.0	396	33.9
Total		815	

(3) Express the result in language that would be used by the experimenter.

(i) The **normal** group scores differ from the **anxious** group scores overall.
(ii) The difference in scores between the two groups depends on which test is used.
(iii) The **one week after** scores differ from the **before drug** scores overall.
(iv) The **anxious** and **normal** groups scores differ 60 minutes after administration of the drug.

(b) Give your views on whether there is a practice effect present and state the statistical test of significance that would help you decide. Do not do any new calculations.

B13.3 An experiment had as aims the investigation of the effect of unilateral auditory stimulation at various levels of intensity and at various frequencies in either the left or right ear on subjects' adjustment of a rod to a position of subjective verticality.

Three intensities were used: 10 dB, 40 dB and 70 dB. Each subject was tested under all six combinations of the **ear** (left, right) and **intensity** conditions. (For analysis purposes **ear** and **intensity** were regarded as a single factor with six levels.)

Subjects were randomly allocated to a single level of the **frequency** condition so that they experienced either 1500 Hz, 1000 Hz or 500 Hz for all six measurements. The **sex** of the subjects was used as a blocking factor, and nine male and nine female subjects were used, making six subjects at each level of **frequency**. The dependent variable was the size of the deviation of the rod from the true vertical.

(a) Set out the data layout diagram for this experiment.
(b) Construct an empty ANOVA summary table for this experiment showing the sources of variance, the degrees of freedom and the appropriate error terms for the denominators of the F-tests.
(c) What features of the experimental conditions would have led the experimenter to choose repeated measures (within-subjects) for one experimental factor and independent groups (between-subjects) for the other?
(d) What might have been the experimenter's reasons for including **sex** as a blocking factor?

FURTHER EXERCISES

1. Ten subjects were asked to decide true or false for a random sample of numeric additions presented on a screen. Additions were sampled from the population consisting of all possible numbers 0 to 9, for example $0+2$, $6+1$, $9+9$ etc.

Half the sums appeared with the correct answer and half with the incorrect answer.

Half the incorrect answers were 'reasonable' (in error by ± 1 or ± 2), and half were 'unreasonable' (in error by ± 5 or ± 6).

Mean reaction times (in milliseconds) were obtained for each subject for choices on the three categories of sums as follows:

Subject	*True*	*False*		*Subject individual mean*
		Reasonable	*Unreasonable*	
1	866	1307	1225	1133
2	1109	1093	992	1065
3	1128	1202	1321	1217
4	1513	1244	1271	1343
5	859	1101	872	944
6	1086	1231	1328	1215
7	1163	1195	1310	1223
8	919	1096	986	1000
9	1068	1119	1117	1101
10	1072	1201	1209	1161
Means	1078.3	1178.9	1163.1	

The SSs were as follows:

Conditions: 58 694
Subjects: 375 792
Conditions × subjects: 231 035

(a) Complete the test of significance of the effect of conditions on reaction time and report the result.
(b) Test an appropriate *a priori* one degree of freedom contrast. Explain why the contrast you have chosen is appropriate. Report the result.
(c) Compare the performances of the fastest and slowest subjects overall by plotting their reaction times on a suitable graph. Interpret the graph for its implications for the analysis of variance and refer to any implications it may have for the designer of the experiment.
(d) Suppose it takes 1 hour to obtain a subject and 0.5 hours to take one subject through one of the three sets of conditions. Compare the efficiency of this design with that of the between-subjects version in which 10 subjects are used for each condition.

2. (a) In independent-groups ANOVA, what are the influences on the power of the test and on the generalizability of the results?
(b) How is it possible, without increasing the cost of the experiment, in a one-factor ANOVA, to improve either the generalizability or the power without worsening the other? Explain the issues.

Appendix C: Solutions to exercises for Chapters 4–13

CHAPTER 4

4.1

(a) Predicted score = 7.88 + 1.15 = 9.03 according to model:

$$\text{Expected score} = 7.88 + \begin{Bmatrix} 0.75 \\ -1.9 \\ 1.15 \end{Bmatrix}$$

(b) $SS_{\text{between}} = 10(0.75^2 + 1.9^2 + 1.15^2)$
$= 10(5.495) = 54.95$

(c) Means are 8.63, 5.98, 9.03 which leads to Fig. C.1.

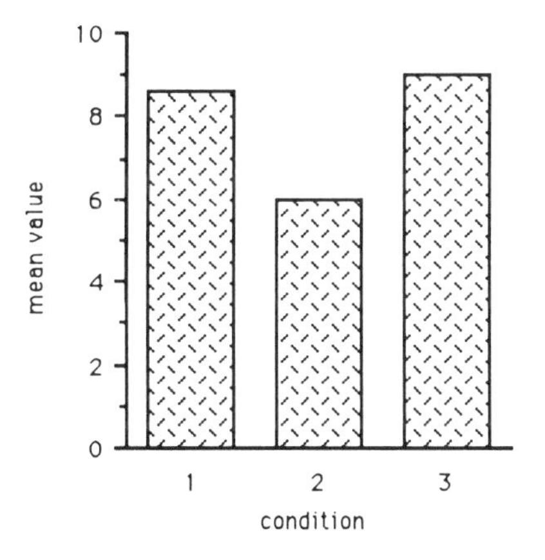

Fig. C.1

4.2 Use an appropriate computer package.
Decide to reject H_0.

4.3

(a) See Fig. C.2.

(b) $\dfrac{\text{size}}{\text{total}} \times 100 = \dfrac{0.2389}{0.2389 + 4.3403} = 5.2\%$ of total

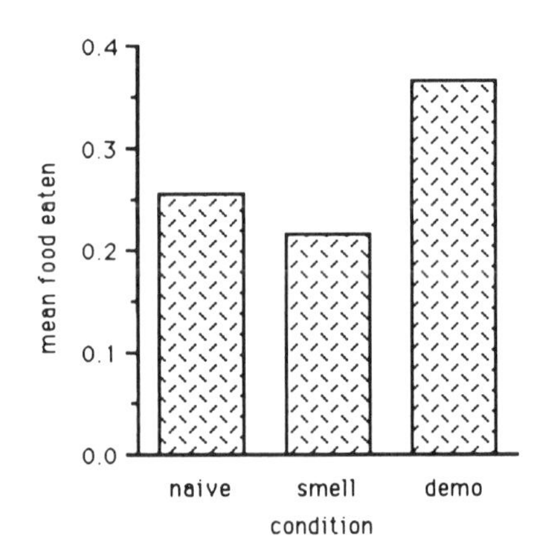

Fig. C.2

(c)

Source	df	SS	MS	F
Between-groups	2	0.2389	0.1195	1.57
Within-groups	57	4.3403	0.076	
Total	59			

F_c (i.e. F critical) on (2, 57) df is 3.15. Since F observed does not exceed 3.15 we cannot reject H_0. Hence we conclude that the treatment conditions do not differ in their effects.

(d) Variance estimate from differences among group means is $MS_{\text{between}} = 0.1195$, whereas variance estimate from difference among individuals within groups is $MS_{\text{within}} = 0.0761$.

If H_0 is true we expect these to be the same. In fact, the between groups estimate is 1.57 times the size of the within groups estimate. This suggests there is a small effect of treatment conditions, but not significant – as seen in (c) above.

CHAPTER 5

5.1

(a) Medication 2 is the more effective (Fig. C.3).

(b) Subject No. 3 appears to have benefitted least from treatment (Fig. C.4).

(c) Overall mean = 15.45
Conditions effect = 7.15, 7.15, −5.65, −8.65
Hence SS for **conditions** = 5(208.99) = 1044.95

(d) Table 1: **conditions** + **subjects** removed
Table 2: **conditions** removed
Table 3: **subjects** removed

(e) SS for reliability $= [0.4^2 + \cdots + (-2.85)^2] = 56.30$

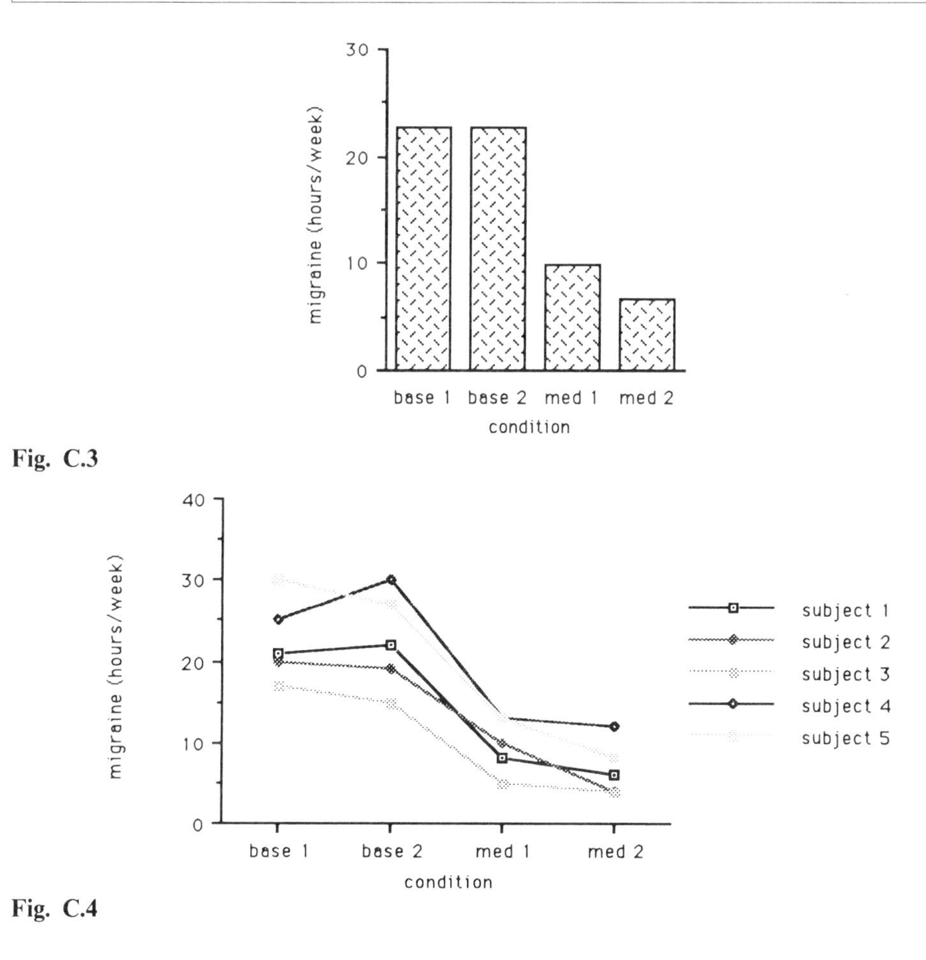

Fig. C.3

Fig. C.4

(f)

Source	*SS*	*df*	*MS*	*F*
Conditions	1044.95	3	348.32	74.5
Subjects	281.70	4		
Reliability	56.30	12	4.68	

F exceeds the critical value. Decide to reject H_0.

(g) Use an appropriate computer package.

(h) Mean square reliability is the required estimate; its value is $56.30/12 = 4.68$.

(i) No. The virtue of this design is that it bypasses the effect of variation between subjects.

5.2

(a) See Fig. C.5. There is a consistent pattern as follows: **drug 3** gives faster reaction time than the other two drugs, *and* all drugs give faster times than the **control** condition.

(b) subject_1 mean is 27.0. Hence the subject_1 **conditions** effect expressed as deviations is:

$$[(30-27)(28-27)(16-27)(34-27)] = [3,\ 1,\ -11,\ 7]$$

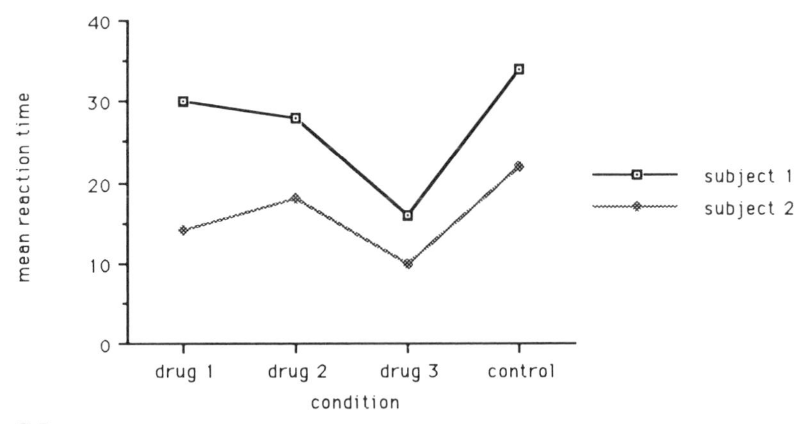

Fig. C.5

and for **subject 2**:

$$[-2, 2, -6, 6]$$

subject 1 is most affected: $SS = 180$
subject 2 is least affected: $SS = \ 80$

(c) **Conditions** as deviations: $[1.5\ 0.7\ -9.3\ 7.1]$. Hence $SS_{\text{conditions}} = 5(1.5^2 + 0.7^2 + \cdots) = 5(139.64) = 698.2$.
Subjects, expressed as deviations:

$$[2.1\ -8.9\ -1.9\ 9.1\ -0.4]$$

Hence $SS_{\text{subjects}} = 4(2.1^2 + \cdots) = 4(170.2) = 680.8$.

(d)

Source	*SS*	*df*	*MS*	*F*
Conditions	698.2	3	232.73	24.76
Subjects	680.8	4		
Reliability	112.8	12	9.4	

Compare with critical F with (3, 12) df:
at 0.05 level = 3.49
at 0.01 level = 5.95
Hence *reject* H_0: **conditions** have identical effects in favour of H_1: **conditions** differ in effects.

(e) Size of SS for **conditions** as a proportion of total:

$$SS_{\text{within}} = \frac{698.2}{698.2 + 112.8} = 0.861$$

Hence 86% of variation within-subjects is explained by **conditions**.

(f) Use an appropriate computer package.

CHAPTER 6

6.1

(a) Best combination is **machine 2**, **method 2**.
Worst combination is **machine 3**, **method 1**.

(b) Overall worst **machine** is No. 3.
Overall worst **training** is No. 4.

(c) See Fig. 6.2. The simple effect of **training** for **machine 3** shows a markedly different pattern from that of the other two simple effects. This suggests that there is an interaction.

(d) Under training **method 4** the simple effect of **machine** is shown in Fig. C.6, whereas the main effect shows a different pattern (Fig. C.7).

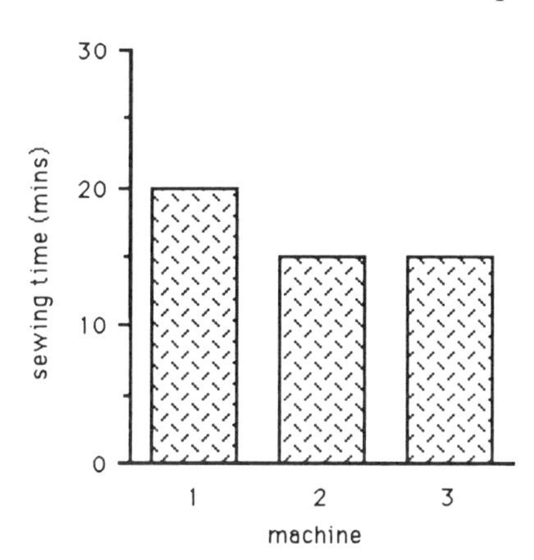

Fig. C.6

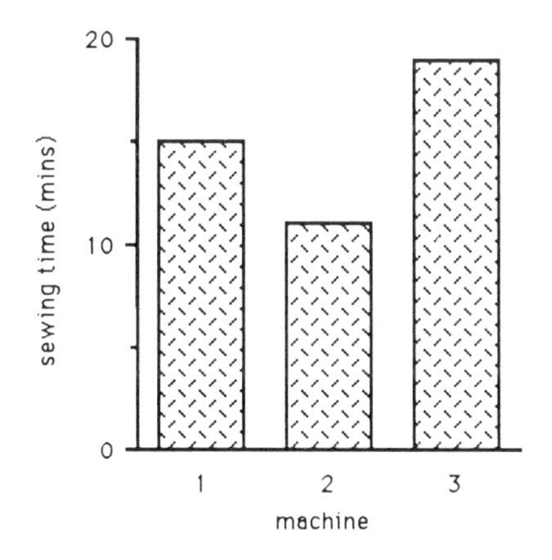

Fig. C.7

(e) Main effect of **machine**:

$$15-15=0$$
$$11-15=-4$$
$$19-15=+4$$

giving (0, -4, $+4$).
Main effect of **training**:

Main effect of **training**:

$15.33-15=0.33$
$12.67-15=-2.33$
$15.33-15=0.33$
$16.67-15=1.67$

giving (0.33, −2.33, 0.33, 1.67).

(f)

$$\begin{Bmatrix} 12 & 13 & 15 & 20 \\ 9 & 8 & 12 & 15 \\ 25 & 17 & 19 & 15 \end{Bmatrix} - \begin{Bmatrix} 0 \\ -4 \\ +4 \end{Bmatrix} = \begin{Bmatrix} 12 & 13 & 15 & 20 \\ 13 & 12 & 16 & 19 \\ 21 & 13 & 15 & 11 \end{Bmatrix}$$

$$\begin{Bmatrix} 12 & 13 & 15 & 20 \\ 13 & 12 & 16 & 19 \\ 21 & 13 & 15 & 11 \end{Bmatrix} - \{0.33 \quad -2.33 \quad 0.33 \quad 1.67\}$$

$$= \begin{Bmatrix} 11.67 & 15.33 & 14.67 & 18.33 \\ 12.67 & 14.33 & 15.67 & 17.33 \\ 20.67 & 15.33 & 14.67 & 9.33 \end{Bmatrix}$$

The remaining variation among the cell means is due to interaction (or to an apparent interaction due to individual variation).

(g) Use an appropriate computer package.

6.2

(a)

Source	*SS*	*df*	*MS*	*F*
Condition	0.048	2	0.024	3.79
Sex	0.096	1	0.096	15.16
Interaction	0.052	2	0.026	4.11
Within (error)	0.342	54	0.00633	

$F_c(2, 54)$ at $0.05=3.23$, $0.01=5.18$ $\quad$ $F_c(1, 54)$ at $0.05=4.08$, $0.01=7.31$

Hence:

condition is significant at $p<0.05$
sex is significant at $p<0.01$
interaction is significant at $p<0.05$

(b) The bar chart showing the effect of factor **sex** is shown in Fig. C.8.
The bar chart showing the effect of factor **condition** is shown in Fig. C.9.
The interaction of **sex** with **condition** is shown in Fig. C.10.

(c) Presence of interaction. Conditions have different effect on females than on males.

(d) By inspection of the appropriate means or calculation of sum of squared deviations for each simple effect. **Sex**: that at demonstration condition (i.e. condition 3)
Condition: that experienced by females

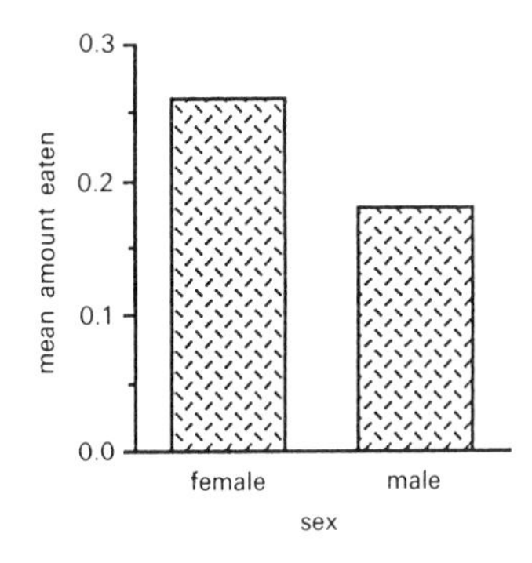

Fig. C.8

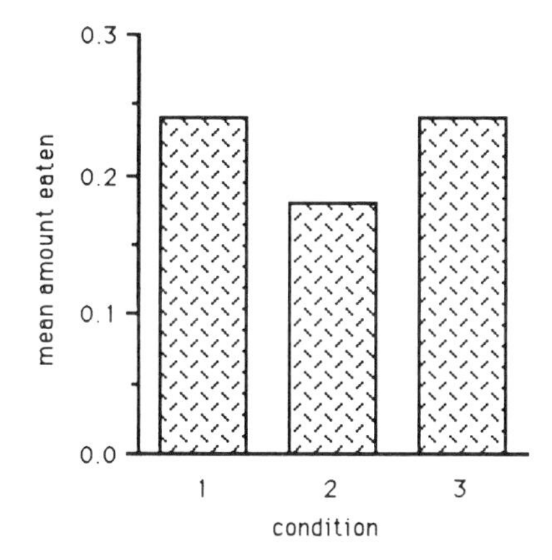

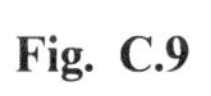
Fig. C.9

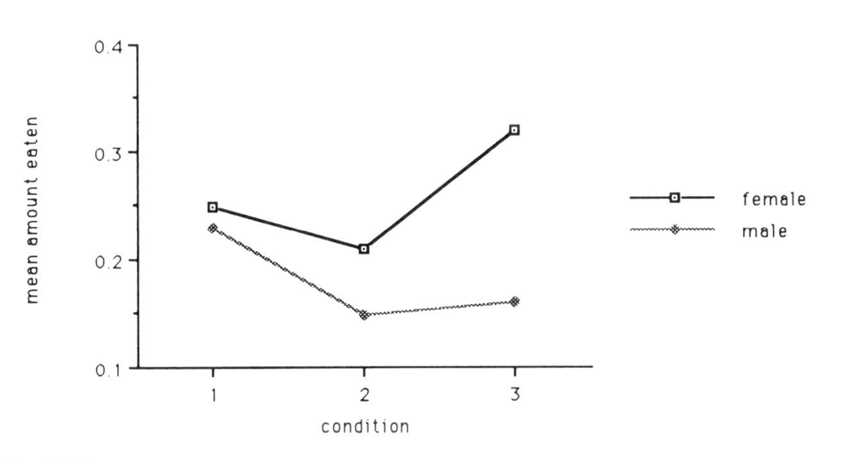

Fig. C.10

(e) Females
Deviations: $(-0.01,\ -0.05,\ +0.06)$

$$SS=10((-0.01)^2+(-0.05)^2+0.06^2)=10(0.0062)$$
$$=0.062$$
$$df=2$$
$$MS=0.031$$
$$F=\frac{0.031}{0.00633}=4.90 \quad \text{on } (2,\ 54)\ df$$

This exceeds the 0.05 critical F, hence reject H_0.

Males
Deviations: $(0.05,\ -0.03,\ -0.02)$

$$SS=10(0.05^2+(-0.03)^2+(-0.02)^2)$$
$$=10(0.0038)$$
$$=0.038$$
$$df=2$$
$$MS=0.019$$
$$F=\frac{0.0109}{0.00633}=3.00 \quad \text{on } (2,\ 54)\ df$$

Not significant, hence do not reject H_0.

(f) If any of the **sex** simple effects are not significant it will be the smallest, i.e. that at the no experience level of **condition** (i.e. condition 1).
Means: 0.25, 0.23
Mean of means:

$$\frac{0.25+0.23}{2}=0.24$$

Deviations: $(+0.01,\ -0.01)$

$$SS=10(0.01^2+(-0.01)^2)=0.002$$
$$df=1$$
$$MS=0.002$$
$$F=\frac{0.002}{0.00633}=0.316$$

Not significant.
This answers the question but note that the simple effect of sex at condition 2 is also not significant.

(g) From (e) the *SS*s of the simple effects of **condition** are 0.062 and 0.038 (total = 0.1).
The *SS*s of the main effect of condition and interaction are 0.048 and 0.052. Total = 0.1 again!
This is a general rule. The *SS*s of the simple effects of a factor at the various levels of another factor total to the *SS*s of the factor itself and its interaction with the other factor!

(h) Females eat more than males (significant at 0.01 level).
'Smell' leads to less food being eaten than the other two conditions (significant at 0.05 level).

Females are more influenced by the conditions than are males; hence there is an interaction ($p<0.05$). (Females eat a larger than expected amount in the 'demo' condition.)

(i) (i) $0.22+(-0.04)+(0.04)+(0.01)=0.15$ g
(ii) $SS=10((-0.03)^2+(-0.01)^2+\cdots+(-0.04)^2)$
$=0.052$

CHAPTER 7

7.1

(a) Unadjusted summary table:

Source	*SS*	*df*	*MS*	*F*
Methods (between groups)	36.95	2	18.475	12.38
Residual (within groups)	26.86	18	1.492	
Total	63.81	20		

F_c on (2, 18) $df=3.55$ at 0.05
$=6.01$ at 0.01

Hence reject H_0.

Adjusted summary table:

Source	SS_{adj}	df_{adj}	*MS*	*F*
Methods (between groups)	16.94	2	8.47	13.98
Residual (within groups)	10.30	17	0.606	
Total	27.24	19		

F_c on (2, 17) $df=3.59$ at 0.05, and 6.11 at 0.01. Hence reject H_0.

(b) Score $=6.24+(-1.33)+0.743(4-2.76)$
$=5.831$

(c) Without covariate, proportion of *SS* is:

$$\text{proportion}=\frac{36.95}{63.81}=57.9\%$$

With covariate partialled out (i.e. adjusted for in the analysis):

$$\text{proportion}=\frac{16.94}{27.24}=62.2\%$$

Thus the **methods** factor looks more impressive when the covariate adjustment is effected.

(d) Use an appropriate computer package.

CHAPTER 8

8.1

To obtain the *MS* for error it is convenient to set out the omnibus ANOVA table:

Source	*df*	*SS*	*MS*
Between-groups	4	5 660.5	1415.1
Within-groups	170	110 939.0	652.6
Total	174		

(a) Choose coefficients -2, -1, 0, $+1$, $+2$ for trend.

$$L=(-2)(9.2)+(-1)(16.8)+(0)(24.0)+(+1)(24.5)+(+2)(21.4)$$
$$=32.1$$

$$SS=\frac{(35)(32.1)^2}{2^2+1^2+1^2+2^2}=3606.435$$

Since *df* is 1, *MS* is 3606.435 too.

$$F \text{ for trend}=\frac{3606.435}{652.6}=5.526$$

Since a directional test is required, obtain F_c as the square of the appropriate t_c (directional).
t_c on 170 *df* at the 0.05 level$=1.658$.
Hence F_c on (1, 170) *df* at 0.05 level (directional)$=(1.658)^2=2.75$.
Decide to reject H_0 at the 0.05 level and conclude that there is a trend in the direction specified.

(b) Choose coefficients -4, $+1$, $+1$, $+1$, $+1$.

$$L=(-4)(9.2)+(+1)(16.8)+(+1)(24.0)+(+1)(24.5)+(+1)(21.4)$$
$$=49.9$$

$$SS=\frac{(35)(49.9)^2}{(-4)^2+1^2+1^2+1^2+1^2}=4357.52$$

Hence

$$F=\frac{4357.52}{652.6}=6.677$$

$$F_c \text{ on } (1, 170)\ df=3.92 \text{ at } 0.05 \text{ level}$$
$$=6.85 \text{ at } 0.01 \text{ level}$$

Decide to reject H_0 at the 0.05 level.
Conclude that the five-minute group differs from the mean of all other groups.

(c) This is a multi-mean test. Therefore the Scheffé adjusted linear contrast is appropriate.

Choose coefficients: -3, -3, $+2$, $+2$, $+2$.

$$L=61.8$$

$$SS=\frac{(35)(61.8)^2}{(-3)^2+(-3)^2+2^2+2^2+2^2}=4455.8$$

$$F=\frac{4455.8}{652.6}=6.83$$

The adjusted F_c is $(k-1)(F_c$ on 4 and 170 $df)$
$F_c=(5-1)(2.45)=\ 9.8$ at 0.05 level
$=(5-1)(3.48)=13.9$ at 0.01 level

Decide not to reject H_0. Conclude that *a posteriori* there is no difference between the means of groups with study times of 10 minutes or less and the others.

(d) Choose a pair-wise test: Newman–Keuls.
Stage 1: set out all means in order of value:

9.2	16.8	21.4	24.0	24.5
(5)	(10)	(25)	(15)	(20)

(5) to (25): two steps apart

Stage 2: the means to be compared are two steps apart. Hence $r=3$, $df=170$.
Stage 3:

$$q=\frac{21.4-9.2}{(652.6/35)^{1/2}}=2.825$$

Stage 4: compare with $q_c=3.36$ at the 0.05 level. Decide not to reject H_0. Conclude that there is no difference between the 5-minute and the 25-minute conditions.

CHAPTER 9

9.1

(a) A: rule of thumb.
Age, when included in the model, has reduced SS by 3 for **conditions** and by 44 for residual. It accounts for 47 out of the total SS of 255. Thus **age** explains

$$\frac{47}{255}=18\%$$

This is less than the criterion 30% and so the decision would be made not to make use of **age**.

B: Compare sensitivities.

Unadjusted for **age**				*Adjusted for* **age**			
Source	*SS*	*df*	*MS*	*Source*	*SS*	*df*	*MS*
Conditions	101	3	33.67	**Conditions**	98	3	32.66
Residual	154	12	12.83	Residual	110	11	10.00

$$\text{Sensitivity} = \frac{4}{12.83} = 0.312 \qquad \text{Sensitivity} = \frac{4}{10.00} = 0.400$$

The version which makes use of **age** as a covariate is more sensitive. Hence decide to include **age** if sensitivity is criterion.

C: Compare efficiencies.

$$\text{Efficiency} = \frac{\text{sensitivity}}{\text{cost}}$$

Cost without use of **age**:

Subjects:	to find 16 @ 0.5 =	8 hours
	to test 16 @ 0.25 =	4 hours
Set-up:		4 hours
	Total	16 hours

Cost with use of **age**: as above but + 3 hours = 19 hours
Hence

efficiency with **age** $= \frac{0.40}{19} = 0.02105$

efficiency without **age** $= \frac{0.312}{16} = 0.0195$

This approach leads to a decision to use **age**.

(b) The relative sensitivity is 0.40/0.312 = 1.282 times as sensitive. The relative efficiency is 0.021 05/0.0195 = 1.08 times as efficient. Hence with method B, comparing sensitivity gives the strongest support for use of **age**.

9.2 Note: **intelligence** scores are assumed to be continuous. However, the experimenter chose to group or block individuals into higher and lower intelligence groups, hence creating a category variable. Also, subjects are randomly allocated from each intelligence block to **techniques**. Hence the name 'Randomized Block Design'.

Blocked				*Continuous covariate*			
Source	*df*	*SS*	*MS*	*Source*	*df*	*SS*	*MS*
Techniques	2	618	309	**Techniques**$_{adj}$	2	590	295
Intelligence	1						
Interaction	2						
Residual	18	4044	224.7	Residual	20	1498	74.9
Total	23			Total$_{adj}$	22		

Blocked:

$$F=\frac{309}{224.7}=1.375$$

F_c on (2, 18) $df=3.55$ at 0.05 level. Hence do not reject H_0.

Continuous covariate:

$$F=\frac{295}{74.9}=3.94$$

F_c on (2, 20) $df=3.49$ at 0.05 level. Hence reject H_0.

When **intelligence** is regarded as a continuous variable, precision is improved, leading to a significant finding.

(b) Note how the 24 subjects are distributed:

		T_1	T_2	T_3
IQ	high	4	4	4
	low	4	4	4

Blocked	*Continuous covariate*
Sensitivity $=\frac{8}{224.7}$	Sensitivity $=\frac{8}{74.9}$
$=0.0356$	$=0.1068$

The costs are identical in the two versions hence:

$$\text{relative efficiency}=\text{relative sensitivity}=\frac{0.1068}{0.0356}=3.0$$

The use of intelligence as a continuous rather than a blocked category variable leads to three times the efficiency.

(c) The only argument in favour of blocking subjects according to intelligence is the easier analysis and availability of interaction from the two-factor ANOVA.

9.3

(a) Use the formula for the approximate method:

$$n=(2)^2(1.96)^2(\text{variance})/\mathbf{spd}^2$$
$$=(15.3664)(\text{variance})/\mathbf{spd}^2$$

Here, variance is estimated as 200. This gives:

$n = (15.3664)(200)/20^2$

$= 8$ subjects per group approximately

(b) Step 1:

The deviations are $(-10, +10)$:

$$t_1^2 + t_2^2 = 200$$

Step 2: Variance is estimated as 200.

Step 3:

Take n to have values: 3, 5, 7, 9, 11, 16, 18. Then the values of ϕ, table values and power are as follows:

Group size (n)	df_e	ϕ	*Table value*	*Power*
3	4	1.22	0.93	0.07
5	8	1.58	0.78	0.22
7	12	1.87	0.60	0.40
9	16	2.12	0.47	0.53
11	20	2.35	0.32	0.68
16	30	2.83	0.13	0.87
18	34	3.00	0.07	0.93

A rough graph for these values is shown in Fig. C.11. The graph describes the relationship of sample size (per group) to power.

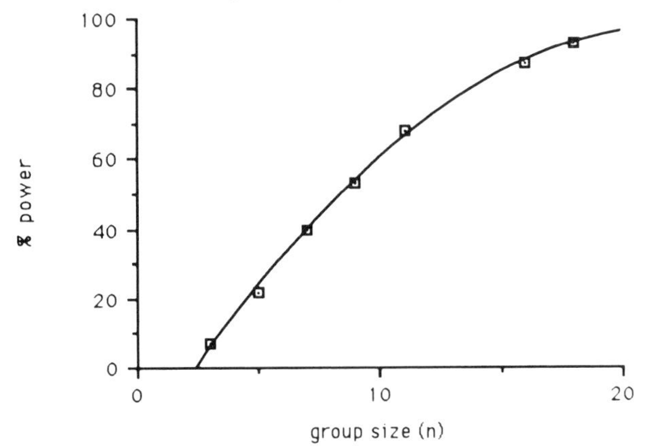

Fig. C.11

(c) Reading from the graph, 8 or 9 subjects per group gives approximately 50% power for the 0.01 significance level.

CHAPTER 10

10.1

(a)
$$r^2 = \frac{83}{255} = 0.325, \quad r = 0.570$$

(b)

$$r^2 = \frac{83+23}{83+23+99} = 0.5171, \quad r = 0.719$$

verbal is 'cut' out of the Venn diagram.

Note that correlation has increased from 0.570.

(c)

$$r^2 = \frac{23+50}{23+50+99} = 0.424, \quad r = 0.651$$

(d)

Unadjusted			*Adjusted*		
Source	*SS*	*df*	*Source*	*SS*	*df*
org	83	1	**org**	106	1
Residual	172	14	Residual	99	13
Total	255	15	Total	205	14
$F = \frac{83}{12.29} = 6.8$			$F = \frac{106}{7.62} = 13.9$		

(e) **Nonverbal** is 'cut' out of the Venn diagram.

$$r^2 = \frac{42}{42+101} = 0.294, \quad r = 0.542$$

Note that correlation has decreased from 0.570.

The inclusion of **verbal** amplifies the effect of **org**, whereas the inclusion of **nonverbal** diminishes the effect of **org** (on the d.v.).

(f)

Unadjusted			*Adjusted*		
Source	*SS*	*df*	*Source*	*SS*	*df*
org	83	1	**org**	42	1
Residual	172	14	Residual	101	13
Total	255	15	Total	143	14
$F = \frac{83}{12.29} = 6.8$			$F = \frac{42}{7.77} = 5.41$		

10.2

(a)

	Male	**Female**	*Adjusted means*	*Unadjusted means*
Traditional	14.00 $n=8$	7.83 $n=6$	10.92	11.36
Imaging	8.80 $n=5$	6.33 $n=9$	7.57	7.21

The effect of adjustment has been to reduce the difference between the effect of the conditions from 4.15 to 3.35. This is reasonable since males return higher scores than females. Males are more frequent under the traditional and females under the imaging method. This **sex** imbalance inflates the apparent treatment effect.

(b)

Unadjusted					*Adjusted*				
Source	*df*	*SS*	*MS*	*F*	*Source*	*df*	*SS*	*MS*	*F*
Method	1	120	120	10.62	**Method**	1	68	68	10.12
Residual	26	295	11.3		Residual	25	168	6.72	
Total	27	415			Total	26	236		

The effect of the adjustment has been to reduce *MS* residual. However, since *MS* for **method** has also been reduced, *F* is unchanged. Thus the significance is unaffected.

10.3

(a)

	Mean Y	*Mean X*
Group **A**	19.4	24.8
Group **B**	32.2	16.4

Plot these on a graph as shown, using + (Fig. C.12). Read off the adjusted mean number of baskets from intersections of fitted lines and 20-month dotted vertical. Estimated adjusted group means:

Group **A**: 10 baskets
Group **B**: 40 baskets

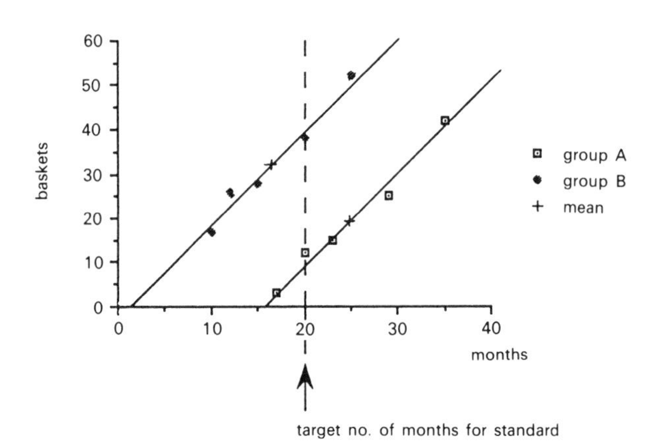

Fig. C.12

(b) See Fig. C.13. Size of effect has been increased from a difference of 12.8 to a difference of approximately 30.

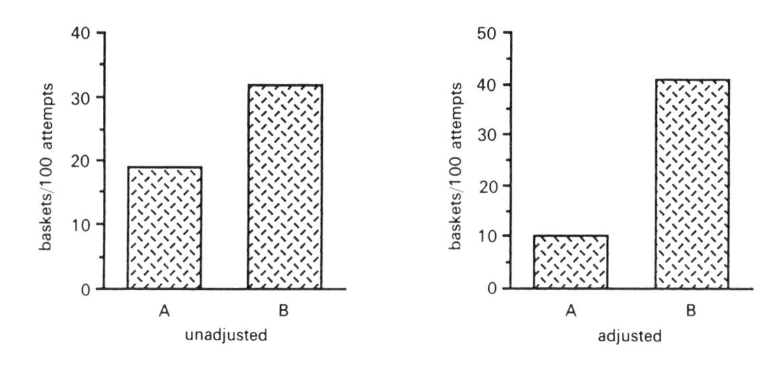

Fig. C.13

(c)

Unadjusted						*Adjusted*					
Source	*df*	*SS*	*MS*	*F*	*p*	*Source*	*df*	*SS*	*MS*	*F*	*p*
Method	1	410	410	2.0	N.S.	**Method**	1	1541	1541	276	<0.001
Residual	8	1598	199			Residual	7	39	5.57		
Total	9	2008				Total	8	1580			

10.4

Investigation 1

(a) i.v.: **height**, continuous, intrinsic
d.v.: **popularity**, continuous
(b) Confounding variables:

age, continuous
ability at sport (poor, average, good) category
parents' income, continuous

(c) Select children to have similar age, same ability at sport, etc. or use adjustment technique.
(d) Not feasible due to i.v. being intrinsic.

Investigation 2

(a) i.v.: **saw programme**, category, intrinsic
d.v.: **vaccinated child**, category

(b) Confounding variables:

social class (category) higher
education (yes/no) category
car owner (yes/no) category

(c) Select mothers according to social class, education etc.
(d) Yes. Sample mothers, randomly allocate to groups to see the video or see a normal child care video. After three weeks check which had child vaccinated.

Investigation 3

(a) i.v.: **type of teaching**, category, extrinsic
d.v.: **score on English test**, continuous
(b) The IQ (continous) or learning ability (poor, average, good) (category) of the pupils may have been higher in one class.
(c) Adjustment techniques at the analysis stage.
(d) Not usually feasible to allocate pupils randomly to a group for more than one day.

Investigation 4

(a) i.v.: **type of conviction**, category, intrinsic
d.v.: **locus of control**, continuous
(b) The burglars may be more intelligent, may have different religious and social class background and differ in age from the violent criminals.
(c) Select for comparison groups of prisoners matched for all these variables.
(d) Not feasible.

CHAPTER 11

11.1

(a) **Final assessment** = 70.0 + 0.35 (**personality** − 49.2) + 1.68 (**aptitude** − 74.6)

(b)

Subject	*Estimated* **final assessment**	*Residual*	*(residual)*^{2}
1	65.2	3.8	14.35
2	57.8	0.2	0.04
3	82.9	−1.9	3.43
4	61.4	−3.4	11.78
5	82.7	1.3	1.66
			31.26

From the Venn diagram it can be seen that the residual SS is given by

$$SS = 606 - 11.27 - 351.21 - 212.27$$
$$= 31.25$$

(c)

Source	*df*	*SS*	*MS*	*F*	*Significance*
Personality	1	362.48	362.48	23.2	*
Aptitude	1	212.27	212.27	13.6	N.S.
Residual	2	31.26	15.63		
Total	4	606.00			

* Significant at the 0.05 level

F_c at 0.05 level is 18.5 on (1, 2) *df*. Decide to reject H_0 for **personality**, to not reject H_0 for **aptitude** (unique).

(d)

$$R^2 = \frac{362.48 + 212.27}{606} = 0.948$$

This means that 94.8% of the total variation is explained by the model.

CHAPTER 12

12.1

(a)

Source	*SS*	*df*	*MS*	*F*
Drugs	32.67	1	32.67	4.460
Tasks	465.08	2	232.54	53.61
Drugs × tasks	11.58	2	5.79	1.335
Subjects	43.95	6	7.325	
Subjects × tasks	52.05	12	4.338	
Total	605.33	23		

F_c on $(1, 6) = 5.99$ at 0.05, 13.8 at 0.01
F_c on $(2, 12) = 3.89$ at 0.05, 6.93 at 0.01
Conclude:

drugs is not significant
tasks is significant, $p < 0.01$
interaction is not significant

(b) The effect of factor **tasks** is shown in Fig. C.14.
The effect of factor **drugs** is shown in Fig. C.15.
The interaction of **tasks** with **drugs** is shown in Fig. C.16.

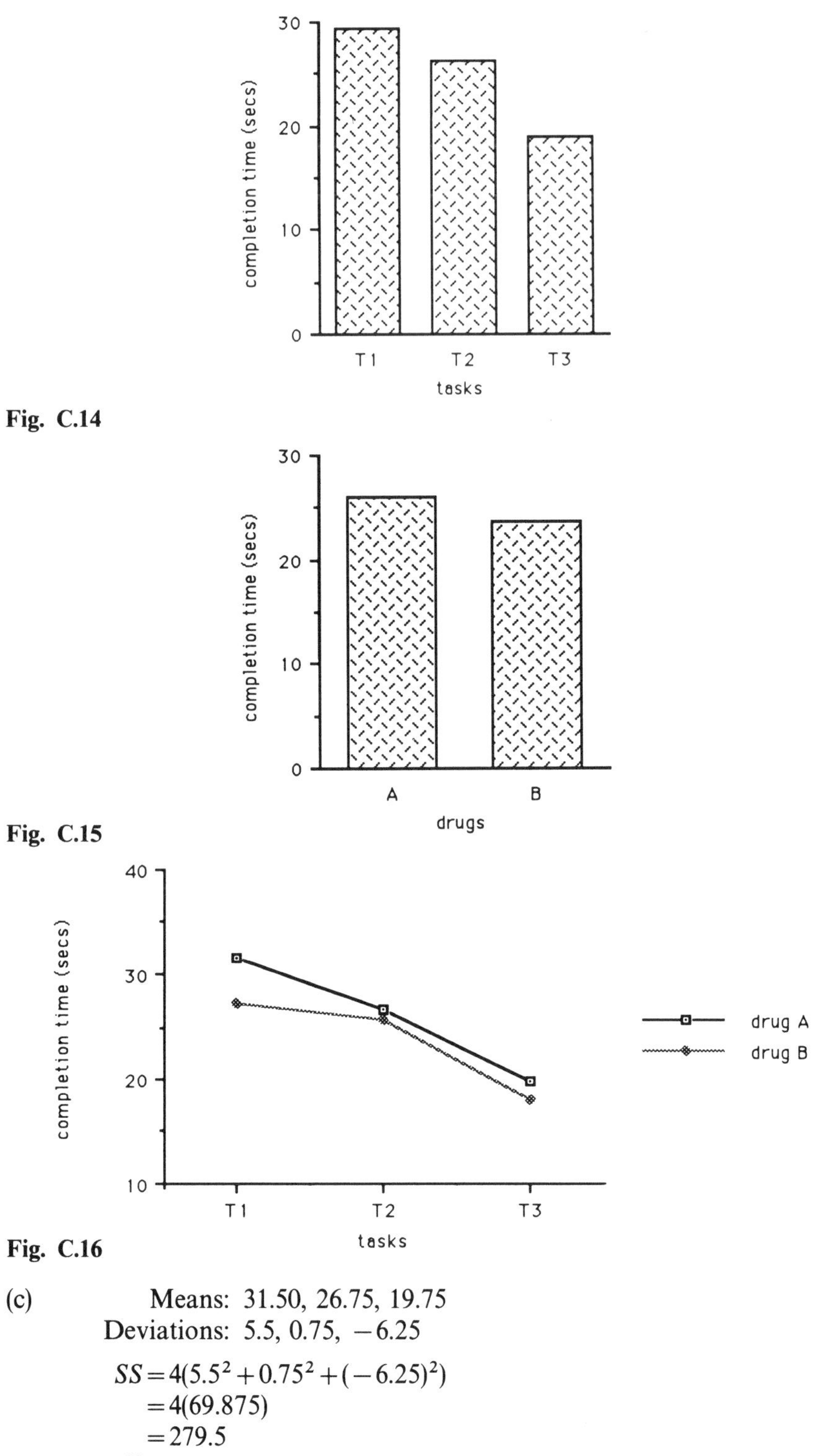

Fig. C.14

Fig. C.15

Fig. C.16

(c) Means: 31.50, 26.75, 19.75

Deviations: 5.5, 0.75, −6.25

$$SS = 4(5.5^2 + 0.75^2 + (-6.25)^2)$$
$$= 4(69.875)$$
$$= 279.5$$
$$df = 2$$
$$MS = 139.75$$

The simple effect is a comparison among the means of the within-subjects factor at a particular level of the between-subjects factor. Therefore (section 12.5.3),

$$MS_{\text{error}} \text{ is } MS_{\text{subjects} \times \text{tasks}} = 4.338$$

So

$$F = \frac{139.75}{4.338} = 32.215 \text{ on } (2, 12)\ df$$

F_c is 6.93 at the 0.01 level, so decide to reject H_0. The simple effect is significant at the 0.01 level.

(d) Coefficients: $-1 \quad +1 \quad 0$

$$L = (-1)(29.375) + (+1)(26.250) = -3.125$$

$$SS = \frac{(8)(3.125)^2}{1+1} = 39.0625$$

$$MS_{\text{error}} \text{ is } MS_{\text{subjects} \times \text{tasks}} = 4.338$$

$$F = \frac{39.0625}{4.338} = 9.005$$

F_c on (1, 12) $df = 4.75$ at 0.05
9.33 at 0.01

Decide to reject H_0. Conclude that task 1 differs from task 2.

(e) The small difference of 2.33 seconds between the mean completion times for drug A and drug B is not significant. Assume that the drugs do not differ in their effects. There is a significant difference between the mean times for the three tasks at the 0.01 level. **Tasks** accounts for 465.08/528.71 = 88% of the within-subject variation in completion times. There is no interaction between **Drugs** and **Tasks**. The simple effect of **Task** for *drug A* is also significant at the 0.01 level.

Task 1 differs from *task 2* at the 0.05 level even though they only differ in mean completion times by 3.125 seconds.

12.2

(a)

Source	*df*	*SS*	*MS*	*F*	*Significance*
L	3	5.70	1.9	2.097	N.S.
D	7	15.63	2.233	3.306	*
L × D	21	14.53	0.6919	1.336	N.S.
Subjects	5	28.34	5.668		
Subjects × L	15	13.59	0.906		
Subjects × D	35	23.64	0.6754		
Subjects × L × D	105	54.35	0.5176		
Total	191				

* Significant at the 0.05 level

F_c (3, 15) df at the 0.05 level $= 3.29$
F_c (7, 35) df at the 0.05 level $= 2.33$
F_c (21, 105) df at the 0.05 level $= 1.75$

(b) (i) Coefficients: $-3, -1, +1, +3$

$$L = -3(1.27) - 1(1.34) + 1(1.40) + 3(1.72)$$
$$= 1.41$$

$$SS = \frac{(8 \times 6)L^2}{\Sigma c^2} = \frac{48(1.41)^2}{20} = 4.771$$

Hence

$$F = \frac{4.771}{0.906} = 5.266$$

The critical F is found as the square of the appropriate critical t, since it is a directional test:

$$t_c\text{(directional) on 15 } df = 1.753 \text{ at 0.05 level}$$
$$= 2.602 \text{ at 0.01 level}$$

This gives $(1.753)^2 = 3.073$ as F_c at 0.05 and $(2.602)^2 = 6.770$ as F_c at 0.01. We decide to reject H_0 at the 0.05 level and conclude that there is a trend in the specified direction.

(ii) Coefficients: $-1, -1, -1, +3$

$$L = -1(1.27) - 1(1.34) - 1(1.40) + 3(1.72)$$
$$= 1.15$$

$$SS = \frac{48(1.15)^2}{1+1+1+9} = 5.29$$

$$F = \frac{5.29}{0.906} = 5.839$$

For *a priori*, F_c (1, 15) $df = 4.54$ at the 0.05 level and 8.68 at the 0.01 level. We reject H_0 at the 5% level (*a priori*).

(c) By selecting on characteristics that influence the dependent variable the power is likely to have been increased.

Right-handedness controls for variation in ability due to handedness. 'Normal' eyesight controls for variation in eyesight. Male and age 20–23, apart from giving a narrow focus to the study, controls for variation in **error distance** between males and females and across age groups.

12.3

Expected score for a randomly selected subject

$$= \underset{\text{Overall mean}}{10.2375} + \underset{\text{Effect of spectacles}}{\begin{Bmatrix} -0.84 \\ +0.84 \end{Bmatrix}} + \underset{\text{Effect of sex}}{\begin{Bmatrix} 0.688 \\ -0.688 \end{Bmatrix}} + \underset{\text{Interaction}}{\begin{Bmatrix} +0.7125 & -0.7125 \\ -0.7125 & +0.7125 \end{Bmatrix}}$$

CHAPTER 13

13.1

(a) The ANOVA summary table is set out below:

Source	*SS*	*df*	*MS*	*Error MS*	*F*	*Significance*
A	2.13	1	2.13	subj.	0.1682	N.S.
subjects (s)	354.6	28	12.66	—	—	
B	55.4	1	55.4	sB	5.875	*
C	4.03	1	4.03	sC	8.550	**
BC	7.75	1	7.75	sBC	30.69	**
AB	0.102	1	0.102	sB	0.011	N.S.
AC	0.833	1	0.833	sC	1.767	N.S.
ABC	0.169	1	0.169	sBC	0.669	N.S.
sB	264.0	28	9.43	—	—	
sC	13.2	28	0.4714	—	—	
sBC	7.07	28	0.2525	—	—	
Total	709.2	119				

F_c on (1, 28) *df* is 4.20 at the 0.05 level and 7.64 at the 0.01 level. Significance is shown as * for the 0.05 level and as ** for the 0.01 level in the ANOVA summary table.

(b) The **speed of movement** has a significant effect on the number of corrections in eye movement (d.v.). The mean number of corrections for the *fast* condition is 4.0125 compared to 2.654 for the *slow* movement.

The **direction of travel** has a significant effect on the number of corrections. The mean for *L to R* is 3.15 corrections compared to 3.517 for *R to L*.

The interaction of **speed of movement** with **direction of travel** has a significant effect on the number of corrections. *Fast* leads to a mean 0.85 more corrections than *slow* in the *L to R* condition, whereas *fast* leads to a mean 1.867 more corrections in the *R to L* condition.

The *dyslexic* do not differ in number of corrections from the *normal* subjects.

No other interactions are significant.

(c) See Fig. C.17.

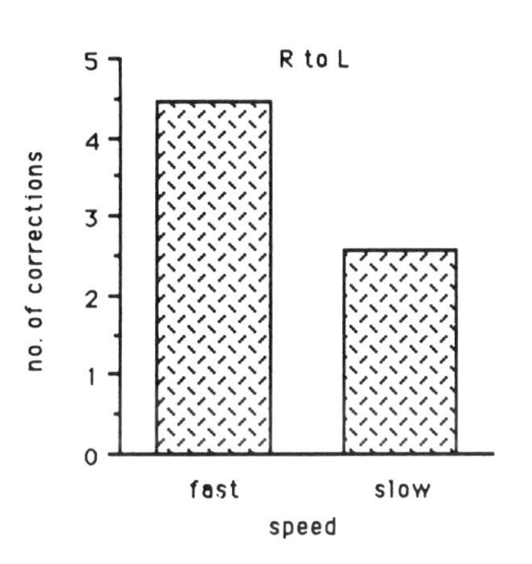

Fig. C.17

13.2

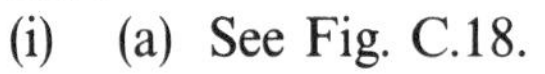

(i) (a) See Fig. C.18.

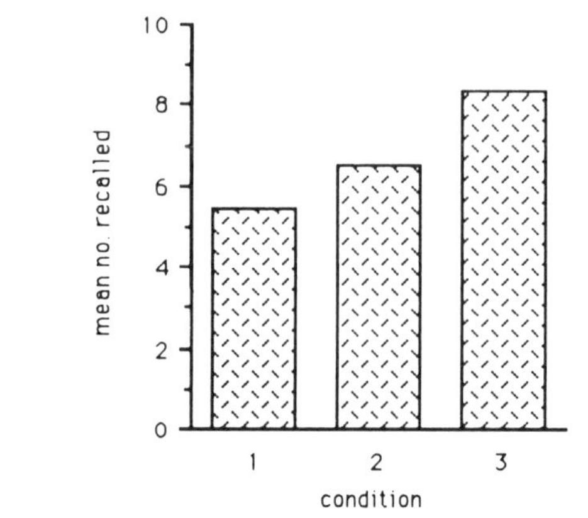

Fig. C.18

(b) SS is 70.8.

Conditions is a between-subjects factor. The total SS between-subjects is 276.9.

Conditions explains $70.8/276.9 = 25.6\%$.

(c) MS_{error} is **subjects**.

$F = 3.68$; F_c at the 0.05 level $= 3.55$, so we reject H_0 and conclude that **conditions** has a significant effect.

(ii) (a) See Fig. C.19.

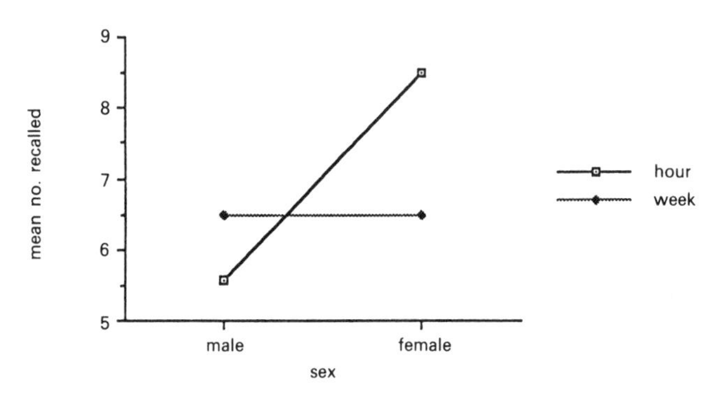

Fig. C.19

(b) $SS = 25.5$.

The total SS within-subjects is 89.4.

Occasion × **Sex** interaction explains 28.5%.

(c) MS_{error} is **subjects** × **occasion**.

$$F = \frac{25.5}{1.95} = 13.08$$

$F_c = 4.41$ at the 0.05 level

$= 8.29$ at the 0.01 level

Reject H_0 at the 0.01 level. Males and females differ in rate of loss of recall.

(iii) (a) See Fig. C.20.

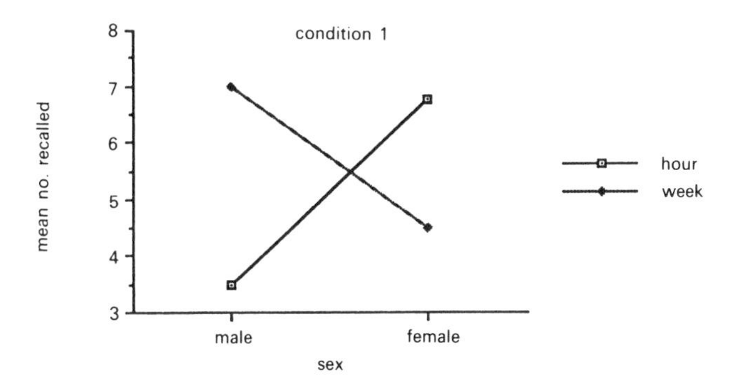

Fig. C.20

(b)

Means observed:

	Hour	*Week*	
Males	3.50	7.00	5.250
Females	6.75	4.50	5.625
	5.125	5.750	5.4375

Means expected:

$$\begin{Bmatrix} 4.9375 & -4.9375 \\ -4.9375 & 4.9375 \end{Bmatrix}$$

Mean deviations:

$$\begin{Bmatrix} -1.4375 & +1.4375 \\ +1.4375 & -1.4375 \end{Bmatrix}$$

$$SS = 4((-1.4375)^2 + 1.4375^2 + 1.4375^2 + (-1.4375)^2)$$
$$= 33.062$$

As a percentage of total SS_{within} this explains 36.95%.

(iv) Simple effect (first-order).
(a) See Fig. C.21.

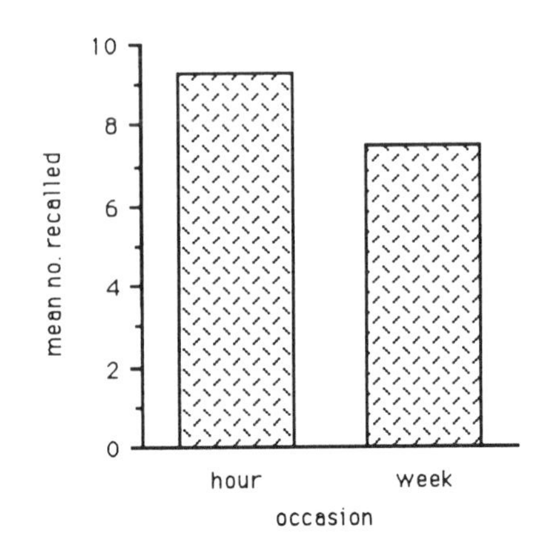

Fig. C.21

(b) Means: (9.25, 7.5); mean of means = 8.375.
Deviations: (0.875, −0.875)

$$SS = 8(0.875^2 + (-0.875)^2)$$
$$= 12.25$$

Expressed as percentage of total $SS_{\text{within}} = 13.69\%$.

Appendix D: Approximate degrees of freedom for test of significance for simple effects in BW and WW designs

D.1 FORMULA FOR DEGREES OF FREEDOM FOR POOLED ERROR FOR BW SIMPLE EFFECTS

$$df_{\text{pooled}} = \frac{(df_{\text{subj}})(df_{\text{subj} \times \text{W}})(\theta + 1)^2}{df_{\text{subj}} + (\theta^2)(df_{\text{subj} \times \text{W}})}$$

where

$$\theta = \frac{SS_{\text{subj}}}{SS_{\text{subj} \times \text{W}}}$$

In the BW example from section 12.2, $\theta = 1166/113 = 10.319$.

$$df_{\text{pooled}} = \frac{(27)(27)(10.319 + 1)^2}{27 + (10.319)^2(27)} = \frac{93399.30}{2902.01} = 32$$

D.2 FORMULA FOR DEGREES OF FREEDOM FOR POOLED ERROR FOR WW SIMPLE EFFECTS

$$df_{\text{pooled}} = \frac{(df_{\text{subj} \times \text{W1}})(df_{\text{subj} \times \text{W2}})(\theta + 1)^2}{df_{\text{subj} \times \text{W1}} + (\theta^2)(df_{\text{subj} \times \text{W2}})}$$

where

$$\theta = \frac{SS_{\text{subj} \times \text{W1}}}{SS_{\text{subj} \times \text{W2}}}$$

Appendix E: Rationale for approximate sample size formula

Consider two conditions whose means differ by $2t$. The equivalent deviations from their common overall mean are $(-t, +t)$. Power as calculated in section 9.4.3 is estimated as follows:

$$\phi = \left(\frac{n(t_1^2 + t_2^2 + t_3^2 + \cdots)}{(k)(\text{variance})} \right)^{1/2}$$

Substitute $k=2$, $t_1 = t_2 = t$, where **spd** $=2t$, and $n=(2)^2(1.96)^2$ (variance)/**spd**^{2} to get $\phi = 1.96$.

Reference to the table of non-central F in Appendix F shows that, for two groups, if the significance level is 0.05 the power values are as follows:

n	5	6	7	9	11	16	inf
df_e	8	10	12	16	20	30	inf
Power	0.68	0.70	0.72	0.74	0.75	0.76	0.79

These values suppose a non-directional test. The power would be greater for a directional test.

Appendix F: Tables of critical values

Table F.1 Student's *t* distribution

	Probability α								
non-directional	*0.60*	*0.50*	*0.40*	*0.25*	*0.20*	*0.10*	***0.05***	*0.02*	***0.01***
directional	*0.30*	*0.25*	*0.20*	*0.125*	*0.10*	***0.05***	*0.025*	***0.01***	*0.005*
df									
1	0.73	1.00	1.38	2.41	3.08	6.31	12.71	31.82	63.66
2	0.62	0.81	1.06	1.60	1.89	2.92	4.30	6.96	9.92
3	0.58	0.79	0.98	1.42	1.64	2.35	3.18	4.54	5.84
4	0.57	0.77	0.94	1.34	1.53	2.13	2.78	3.75	4.60
5	0.56	0.75	0.92	1.30	1.48	2.02	2.57	3.36	4.03
6	0.55	0.74	0.91	1.27	1.44	1.94	2.45	3.14	3.71
7	0.55	0.73	0.90	1.25	1.42	1.89	2.36	3.00	3.50
8	0.55	0.72	0.89	1.24	1.40	1.86	2.31	2.90	3.36
9	0.54	0.71	0.88	1.23	1.38	1.83	2.26	2.82	3.25
10	0.54	0.70	0.88	1.22	1.37	1.81	2.23	2.76	3.17
11	0.54	0.70	0.88	1.21	1.36	1.80	2.20	2.72	3.11
12	0.54	0.69	0.87	1.21	1.36	1.78	2.18	2.68	3.05
13	0.54	0.69	0.87	1.21	1.35	1.77	2.16	2.65	3.01
14	0.54	0.69	0.87	1.20	1.34	1.76	2.14	2.62	2.98
15	0.54	0.69	0.87	1.20	1.34	1.75	2.13	2.60	2.95
16	0.54	0.69	0.86	1.20	1.34	1.75	2.12	2.58	2.92
17	0.53	0.69	0.86	1.19	1.33	1.74	2.11	2.57	2.90
18	0.53	0.69	0.86	1.19	1.33	1.73	2.10	2.55	2.88
19	0.53	0.69	0.86	1.19	1.33	1.73	2.09	2.54	2.86
20	0.53	0.69	0.86	1.18	1.32	1.72	2.09	2.53	2.85
21	0.53	0.69	0.86	1.18	1.32	1.72	2.08	2.52	2.83
22	0.53	0.69	0.86	1.18	1.32	1.72	2.07	2.51	2.82
23	0.53	0.68	0.86	1.18	1.32	1.71	2.07	2.50	2.81
24	0.53	0.68	0.86	1.18	1.32	1.71	2.06	2.49	2.80
25	0.53	0.68	0.86	1.18	1.32	1.71	2.06	2.49	2.79
26	0.53	0.68	0.86	1.18	1.32	1.71	2.06	2.48	2.78
27	0.53	0.68	0.86	1.18	1.31	1.70	2.05	2.47	2.77
28	0.53	0.68	0.86	1.17	1.31	1.70	2.05	2.47	2.76
29	0.53	0.68	0.85	1.17	1.31	1.70	2.05	2.46	2.76
30	0.53	0.68	0.85	1.17	1.31	1.70	2.04	2.46	2.75
40	0.53	0.68	0.85	1.17	1.30	1.68	2.02	2.42	2.70
50	0.53	0.67	0.85	1.16	1.30	1.68	2.01	2.40	2.68
60	0.53	0.67	0.85	1.16	1.30	1.67	2.00	2.39	2.66
80	0.53	0.67	0.85	1.16	1.29	1.66	1.99	2.37	2.64
100	0.53	0.67	0.84	1.16	1.29	1.66	1.98	2.36	2.63
200	0.52	0.67	0.84	1.15	1.29	1.65	1.97	2.35	2.60
300	0.52	0.67	0.84	1.15	1.28	1.65	1.96	2.33	2.59
∞	0.52	0.67	0.84	1.15	1.28	1.64	1.96	2.33	2.58

Table F.2 Critical values of the *F*-distribution

df for denominator	α_1	α_2	*df for numerator* 1	2	3	4	5	6	7	8	9	10	12	15	20	24	30	40	60	∞
	0.125	0.25	5.83	7.50	8.20	8.58	8.82	8.98	9.10	9.19	9.26	9.32	9.41	9.49	9.58	9.63	9.67	9.71	9.76	9.85
	0.05	0.10	39.9	49.5	53.6	55.8	57.2	58.2	58.9	59.4	59.9	60.2	60.7	61.2	61.7	62.0	62.3	62.5	62.8	63.3
	0.025	**0.05**	**161**	**200**	**216**	**225**	**230**	**234**	**237**	**239**	**240**	**242**	**244**	**246**	**248**	**249**	**250**	**251**	**252**	**254**
1	0.0125	0.025	648	800	864	900	922	937	948	957	963	969	977	985	993	997	1001	1006	1010	1018
	0.005	**0.01**	**4052**	**5000**	**5403**	**5625**	**5764**	**5859**	**5928**	**5982**	**6022**	**6056**	**6106**	**6157**	**6209**	**6235**	**6261**	**6287**	**6313**	**6366**
	0.0005	0.001	4053*	5000*	5404*	5625*	5764*	5859*	5929*	5981*	6023*	6056*	6107*	6158*	6209*	6235*	6261*	6287*	6313*	6366*
	0.125	0.25	2.57	3.00	3.15	3.23	3.28	3.31	3.34	3.35	3.37	3.38	3.39	3.41	3.43	3.43	3.44	3.45	3.46	3.48
	0.05	0.10	8.53	9.00	9.16	9.24	9.29	9.33	9.35	9.37	9.38	9.39	9.41	9.42	9.44	9.45	9.46	9.47	9.47	9.49
	0.025	**0.05**	**18.5**	**19.0**	**19.2**	**19.3**	**19.3**	**19.3**	**19.4**	**19.4**	**19.4**	**19.4**	**19.4**	**19.4**	**19.5**	**19.5**	**19.5**	**19.5**	**19.5**	**19.5**
2	0.0125	0.025	38.5	39.0	39.2	39.3	39.3	39.3	39.4	39.4	39.4	39.4	39.4	39.4	39.5	39.5	39.5	39.5	39.5	39.5
	0.005	**0.01**	**98.5**	**99.0**	**99.2**	**99.3**	**99.3**	**99.3**	**99.4**	**99.4**	**99.4**	**99.4**	**99.4**	**99.4**	**99.5**	**99.5**	**99.5**	**99.5**	**99.5**	**99.5**
	0.0005	0.001	999	999	999	999	999	999	999	999	999	999	999	999	999	1000	1000	1000	1000	1000
	0.125	0.25	2.02	2.28	2.36	2.39	2.41	2.42	2.43	2.44	2.44	2.44	2.45	2.46	2.46	2.46	2.47	2.47	2.47	2.47
	0.05	0.10	5.54	5.46	5.39	5.34	5.31	5.28	5.27	5.25	5.24	5.23	5.22	5.20	5.18	5.18	5.17	5.16	5.15	5.13
	0.025	**0.05**	**10.1**	**9.55**	**9.28**	**9.12**	**9.01**	**8.94**	**8.89**	**8.85**	**8.81**	**8.79**	**8.74**	**8.70**	**8.66**	**8.64**	**8.62**	**8.59**	**8.57**	**8.53**
3	0.0125	0.025	17.4	16.0	15.4	15.1	14.9	14.7	14.6	14.5	14.5	14.4	14.3	14.2	14.2	14.1	14.1	14.0	14.0	13.9
	0.005	**0.01**	**34.1**	**30.8**	**29.5**	**28.7**	**28.2**	**27.9**	**27.7**	**27.5**	**27.4**	**27.2**	**27.0**	**26.9**	**26.7**	**26.6**	**26.5**	**26.4**	**26.3**	**26.1**
	0.0005	0.001	167	148	141	137	135	133	132	131	130	129	128	127	126	126	125	125	124	124
	0.125	0.25	1.81	2.00	2.05	2.06	2.07	2.08	2.08	2.08	2.08	2.08	2.08	2.08	2.08	2.08	2.08	2.08	2.08	2.08
	0.05	0.10	4.54	4.32	4.19	4.11	4.05	4.01	3.98	3.95	3.94	3.92	3.90	3.87	3.84	3.83	3.82	3.80	3.79	3.76
	0.025	**0.05**	**7.71**	**6.94**	**6.59**	**6.39**	**6.26**	**6.16**	**6.09**	**6.04**	**6.00**	**5.96**	**5.91**	**5.86**	**5.80**	**5.77**	**5.75**	**5.72**	**5.69**	**5.63**
4	0.0125	0.025	12.2	10.6	9.98	9.60	9.36	9.20	9.07	8.98	8.90	8.84	8.75	8.66	8.56	8.51	8.46	8.41	8.36	8.26
	0.005	**0.01**	**21.2**	**18.0**	**16.7**	**16.0**	**15.5**	**15.2**	**15.0**	**14.8**	**14.7**	**14.6**	**14.4**	**14.2**	**14.0**	**13.9**	**13.8**	**13.8**	**13.6**	**13.5**
	0.0005	0.001	74.1	61.2	56.2	53.4	51.7	50.5	49.7	49.0	48.5	48.0	47.4	46.8	46.1	45.8	45.4	45.1	44.8	44.0

* These values must be multiplied by 100.
α_1 and α_2 are the probabilities corresponding to directional and non-directional tests respectively.

df for denominator	α_1	α_2	1	2	3	4	5	6	7	8	9	10	12	15	20	24	30	40	60	∞
	0.125	0.25	1.69	1.85	1.88	1.89	1.89	1.89	1.89	1.89	1.89	1.89	1.89	1.89	1.88	1.88	1.88	1.88	1.87	1.87
	0.05	0.10	4.06	3.78	3.62	3.52	3.45	3.40	3.37	3.34	3.32	3.30	3.27	3.24	3.21	3.19	3.17	3.16	3.14	3.10
	0.025	**0.05**	**6.61**	**5.79**	**5.41**	**5.19**	**5.05**	**4.95**	**4.88**	**4.82**	**4.77**	**4.74**	**4.68**	**4.62**	**4.56**	**4.53**	**4.50**	**4.46**	**4.43**	**4.36**
5	0.0125	0.025	10.0	8.43	7.76	7.39	7.15	6.98	6.85	6.76	6.68	6.62	6.52	6.43	6.33	6.28	6.23	6.18	6.12	6.02
	0.005	**0.01**	**16.3**	**13.3**	**12.1**	**11.4**	**11.0**	**10.7**	**10.5**	**10.3**	**10.2**	**10.0**	**9.89**	**9.72**	**9.55**	**9.47**	**9.38**	**9.29**	**9.20**	**9.02**
	0.0005	0.001	47.2	37.1	33.2	31.1	29.8	28.8	28.2	27.6	27.2	26.9	26.4	25.9	25.4	25.1	24.9	24.6	24.3	23.8
	0.125	0.25	1.62	1.76	1.78	1.79	1.79	1.78	1.78	1.78	1.77	1.77	1.77	1.76	1.76	1.75	1.75	1.75	1.74	1.74
	0.05	0.10	3.78	3.46	3.29	3.18	3.11	3.05	3.01	2.98	2.96	2.94	2.90	2.87	2.84	2.82	2.80	2.78	2.76	2.72
	0.025	**0.05**	**5.99**	**5.14**	**4.76**	**4.53**	**4.39**	**4.28**	**4.21**	**4.15**	**4.10**	**4.06**	**4.00**	**3.94**	**3.87**	**3.84**	**3.81**	**3.77**	**3.74**	**3.67**
6	0.0125	0.025	8.81	7.26	6.60	6.23	5.99	5.82	5.70	5.60	5.52	5.46	5.37	5.27	5.17	5.12	5.07	5.01	4.96	4.85
	0.005	**0.01**	**13.8**	**10.9**	**9.78**	**9.15**	**8.75**	**8.47**	**8.26**	**8.10**	**7.98**	**7.87**	**7.72**	**7.56**	**7.40**	**7.31**	**7.23**	**7.14**	**7.06**	**6.88**
	0.0005	0.001	35.5	27.0	23.7	21.9	20.8	20.0	19.5	19.0	18.7	18.4	18.0	17.6	17.1	16.9	16.7	16.4	16.2	15.8
	0.125	0.25	1.57	1.70	1.72	1.72	1.71	1.71	1.70	1.70	1.69	1.69	1.68	1.68	1.67	1.67	1.66	1.66	1.65	1.65
	0.05	0.10	3.59	3.26	3.07	2.96	2.88	2.83	2.78	2.75	2.72	2.70	2.67	2.63	2.59	2.58	2.56	2.54	2.51	2.47
	0.025	**0.05**	**5.59**	**4.74**	**4.35**	**4.12**	**3.97**	**3.87**	**3.79**	**3.73**	**3.68**	**3.64**	**3.57**	**3.51**	**3.44**	**3.41**	**3.38**	**3.34**	**3.30**	**3.23**
7	0.0125	0.025	8.07	6.54	5.89	5.52	5.29	5.12	4.99	4.90	4.82	4.76	4.67	4.57	4.47	4.42	4.36	4.31	4.25	4.14
	0.005	**0.01**	**12.2**	**9.55**	**8.45**	**7.85**	**7.46**	**7.19**	**6.99**	**6.84**	**6.72**	**6.62**	**6.47**	**6.31**	**6.16**	**6.07**	**5.99**	**5.91**	**5.82**	**5.65**
	0.0005	0.001	29.2	21.7	18.8	17.2	16.2	15.5	15.0	14.6	14.3	14.1	13.7	13.3	12.9	12.7	12.5	12.3	12.1	11.7
	0.125	0.25	1.54	1.66	1.67	1.66	1.66	1.65	1.64	1.64	1.63	1.63	1.62	1.62	1.61	1.60	1.60	1.59	1.59	1.58
	0.05	0.10	3.46	3.11	2.92	2.81	2.73	2.67	2.62	2.59	2.56	2.54	2.50	2.46	2.42	2.40	2.38	2.36	2.34	2.29
	0.025	**0.05**	**5.32**	**4.46**	**4.07**	**3.84**	**3.69**	**3.58**	**3.50**	**3.44**	**3.39**	**3.35**	**3.28**	**3.22**	**3.15**	**3.12**	**3.08**	**3.04**	**3.01**	**2.93**
8	0.0125	0.025	7.57	6.06	5.42	5.05	4.82	4.65	4.53	4.43	4.36	4.30	4.20	4.10	4.00	3.95	3.89	3.84	3.78	3.67
	0.005	**0.01**	**11.3**	**8.65**	**7.59**	**7.01**	**6.63**	**6.37**	**6.18**	**6.03**	**5.91**	**5.81**	**5.67**	**5.52**	**5.36**	**5.28**	**5.20**	**5.12**	**5.03**	**4.86**
	0.0005	0.001	25.4	18.5	15.8	14.4	13.5	12.9	12.4	12.0	11.8	11.5	11.2	10.8	10.5	10.3	10.1	9.92	9.73	9.33

df for denominator	α_1	α_2	*df for numerator* **1**	**2**	**3**	**4**	**5**	**6**	**7**	**8**	**9**	**10**	**12**	**15**	**20**	**24**	**30**	**40**	**60**	∞
	0.125	0.25	1.51	1.62	1.63	1.63	1.62	1.61	1.60	1.60	1.59	1.59	1.58	1.57	1.56	1.56	1.55	1.54	1.54	1.53
	0.05	0.10	3.36	3.01	2.81	2.69	2.61	2.55	2.51	2.47	2.44	2.42	2.38	2.34	2.30	2.28	2.25	2.23	2.21	2.16
	0.025	**0.05**	**5.12**	**4.26**	**3.86**	**3.63**	**3.48**	**3.37**	**3.29**	**3.23**	**3.18**	**3.14**	**3.07**	**3.01**	**2.94**	**2.90**	**2.86**	**2.83**	**2.79**	**2.71**
9	0.0125	0.025	7.21	5.71	5.08	4.72	4.48	4.32	4.20	4.10	4.03	3.96	3.87	3.77	3.67	3.61	3.56	3.51	3.45	3.33
	0.005	**0.01**	**10.6**	**8.02**	**6.99**	**6.42**	**6.06**	**5.80**	**5.61**	**5.47**	**5.35**	**5.26**	**5.11**	**4.96**	**4.81**	**4.73**	**4.65**	**4.57**	**4.48**	**4.31**
	0.0005	0.001	22.9	16.4	13.9	12.6	11.7	11.1	10.7	10.4	10.1	9.89	9.57	9.24	8.90	8.72	8.55	8.37	8.19	7.81
	0.125	0.25	1.49	1.60	1.60	1.59	1.59	1.58	1.57	1.56	1.56	1.55	1.54	1.53	1.52	1.52	1.51	1.51	1.50	1.48
	0.05	0.10	3.29	2.92	2.73	2.61	2.52	2.46	2.41	2.38	2.35	2.32	2.28	2.24	2.20	2.18	2.16	2.13	2.11	2.06
	0.025	**0.05**	**4.96**	**4.10**	**3.71**	**3.48**	**3.33**	**3.22**	**3.14**	**3.07**	**3.02**	**2.98**	**2.91**	**2.85**	**2.77**	**2.74**	**2.70**	**2.66**	**2.62**	**2.54**
10	0.0125	0.025	6.94	5.46	4.83	4.47	4.24	4.07	3.95	3.85	3.78	3.72	3.62	3.52	3.42	3.37	3.31	3.26	3.20	3.08
	0.005	**0.01**	**10.0**	**7.56**	**6.55**	**5.99**	**5.64**	**5.39**	**5.20**	**5.06**	**4.94**	**4.85**	**4.71**	**4.56**	**4.41**	**4.33**	**4.25**	**4.17**	**4.08**	**3.91**
	0.0005	0.001	21.0	14.9	12.6	11.3	10.5	9.92	9.52	9.20	8.96	8.75	8.45	8.13	7.80	7.64	7.47	7.30	7.12	6.76
	0.125	0.25	1.47	1.58	1.58	1.57	1.56	1.55	1.54	1.53	1.53	1.52	1.51	1.50	1.49	1.49	1.48	1.47	1.47	1.45
	0.05	0.10	3.23	2.86	2.66	2.54	2.45	2.39	2.34	2.30	2.27	2.25	2.21	2.17	2.12	2.10	2.08	2.05	2.03	1.97
	0.025	**0.05**	**4.84**	**3.98**	**3.59**	**3.36**	**3.20**	**3.09**	**3.01**	**2.95**	**2.90**	**2.85**	**2.79**	**2.72**	**2.65**	**2.61**	**2.57**	**2.53**	**2.49**	**2.40**
11	0.0125	0.025	6.72	5.26	4.63	4.28	4.04	3.88	3.76	3.66	3.59	3.53	3.43	3.33	3.23	3.17	3.12	3.06	3.00	2.88
	0.005	**0.01**	**9.65**	**7.21**	**6.22**	**5.67**	**5.32**	**5.07**	**4.89**	**4.74**	**4.63**	**4.54**	**4.40**	**4.25**	**4.10**	**4.02**	**3.94**	**3.86**	**3.78**	**3.60**
	0.0005	0.001	19.7	13.8	11.6	10.4	9.58	9.05	8.66	8.35	8.12	7.92	7.63	7.32	7.01	6.85	6.68	6.52	6.35	6.00
	0.125	0.25	1.46	1.56	1.56	1.55	1.54	1.53	1.52	1.51	1.51	1.50	1.49	1.48	1.47	1.46	1.45	1.45	1.44	1.42
	0.05	0.10	3.18	2.81	2.61	2.48	2.39	2.33	2.28	2.24	2.21	2.19	2.15	2.10	2.06	2.04	2.01	1.99	1.96	1.90
	0.025	**0.05**	**4.75**	**3.89**	**3.49**	**3.26**	**3.11**	**3.00**	**2.91**	**2.85**	**2.80**	**2.75**	**2.69**	**2.62**	**2.54**	**2.51**	**2.47**	**2.43**	**2.38**	**2.30**
12	0.0125	0.025	6.55	5.10	4.47	4.12	3.89	3.73	3.61	3.51	3.44	3.37	3.28	3.18	3.07	3.02	2.96	2.91	2.85	2.72
	0.005	**0.01**	**9.33**	**6.93**	**5.95**	**5.41**	**5.06**	**4.82**	**4.64**	**4.50**	**4.39**	**4.30**	**4.16**	**4.01**	**3.86**	**3.78**	**3.70**	**3.62**	**3.54**	**3.36**
	0.0005	0.001	18.6	13.0	10.8	9.63	8.89	8.38	8.00	7.71	7.48	7.29	7.00	6.71	6.40	6.25	6.09	5.93	5.76	5.42
	0.125	0.25	1.45	1.55	1.55	1.53	1.52	1.51	1.50	1.49	1.49	1.48	1.47	1.46	1.45	1.44	1.43	1.42	1.42	1.40
	0.05	0.10	3.14	2.76	2.56	2.43	2.35	2.28	2.23	2.20	2.16	2.14	2.10	2.05	2.01	1.98	1.96	1.93	1.90	1.85
	0.025	**0.05**	**4.67**	**3.81**	**3.41**	**3.18**	**3.03**	**2.92**	**2.83**	**2.77**	**2.71**	**2.67**	**2.60**	**2.53**	**2.46**	**2.42**	**2.38**	**2.34**	**2.30**	**2.21**
13	0.0125	0.025	6.41	4.97	4.35	4.00	3.77	3.60	3.48	3.39	3.31	3.25	3.15	3.05	2.95	2.89	2.84	2.78	2.72	2.60
	0.005	**0.01**	**9.07**	**6.70**	**5.74**	**5.21**	**4.86**	**4.62**	**4.44**	**4.30**	**4.19**	**4.10**	**3.96**	**3.82**	**3.66**	**3.59**	**3.51**	**3.43**	**3.34**	**3.17**
	0.0005	0.001	17.8	12.3	10.2	9.07	8.35	7.86	7.49	7.21	6.98	6.80	6.52	6.23	5.93	5.78	5.63	5.47	5.30	4.97

df for denominator	α_1	α_2	*df for numerator* 1	2	3	4	5	6	7	8	9	10	12	15	20	24	30	40	60	∞
	0.125	0.25	1.44	1.53	1.53	1.52	1.51	1.50	1.49	1.48	1.47	1.46	1.45	1.44	1.43	1.42	1.41	1.41	1.40	1.38
	0.05	0.10	3.10	2.73	2.52	2.39	2.31	2.24	2.19	2.15	2.12	2.10	2.05	2.01	1.96	1.94	1.91	1.89	1.86	1.80
	0.025	**0.05**	**4.60**	**3.74**	**3.34**	**3.11**	**2.96**	**2.85**	**2.76**	**2.70**	**2.65**	**2.60**	**2.53**	**2.46**	**2.39**	**2.35**	**2.31**	**2.27**	**2.22**	**2.13**
14	0.0125	0.025	6.30	4.86	4.24	3.89	3.66	3.50	3.38	3.29	3.21	3.15	3.05	2.95	2.84	2.79	2.73	2.67	2.61	2.49
	0.005	**0.01**	**8.86**	**6.51**	**5.56**	**5.04**	**4.69**	**4.46**	**4.28**	**4.14**	**4.03**	**3.94**	**3.80**	**3.66**	**3.51**	**3.43**	**3.35**	**3.27**	**3.18**	**3.00**
	0.0005	0.001	17.1	11.8	9.73	8.62	7.92	7.43	7.08	6.80	6.58	6.40	6.13	5.85	5.56	5.41	5.25	5.10	4.94	4.60
	0.125	0.25	1.43	1.52	1.52	1.51	1.49	1.48	1.47	1.46	1.46	1.45	1.44	1.43	1.41	1.41	1.40	1.39	1.38	1.36
	0.05	0.10	3.07	2.70	2.49	2.36	2.27	2.21	2.16	2.12	2.09	2.06	2.02	1.97	1.92	1.90	1.87	1.85	1.82	1.76
	0.025	**0.05**	**4.54**	**3.68**	**3.29**	**3.06**	**2.90**	**2.79**	**2.71**	**2.64**	**2.59**	**2.54**	**2.48**	**2.40**	**2.33**	**2.29**	**2.25**	**2.20**	**2.16**	**2.07**
15	0.0125	0.025	6.20	4.77	4.15	3.80	3.58	3.41	3.29	3.20	3.12	3.06	2.96	2.86	2.76	2.70	2.64	2.59	2.52	2.40
	0.005	**0.01**	**8.68**	**6.36**	**5.42**	**4.89**	**4.56**	**4.32**	**4.14**	**4.00**	**3.89**	**3.80**	**3.67**	**3.52**	**3.37**	**3.29**	**3.21**	**3.13**	**3.05**	**2.87**
	0.0005	0.001	16.6	11.3	9.34	8.25	7.57	7.09	6.74	6.47	6.26	6.08	5.81	5.54	5.25	5.10	4.95	4.80	4.64	4.31
	0.125	0.25	1.42	1.51	1.51	1.50	1.48	1.47	1.46	1.45	1.44	1.44	1.43	1.41	1.40	1.39	1.38	1.37	1.36	1.34
	0.05	0.10	3.05	2.67	2.46	2.33	2.24	2.18	2.13	2.09	2.06	2.03	1.99	1.94	1.89	1.87	1.84	1.81	1.78	1.72
	0.025	**0.05**	**4.49**	**3.63**	**3.24**	**3.01**	**2.85**	**2.74**	**2.66**	**2.59**	**2.54**	**2.49**	**2.42**	**2.35**	**2.28**	**2.24**	**2.19**	**2.15**	**2.11**	**2.01**
16	0.0125	0.025	6.12	4.69	4.08	3.73	3.50	3.34	3.22	3.12	3.05	2.99	2.89	2.79	2.68	2.63	2.57	2.51	2.45	2.32
	0.005	**0.01**	**8.53**	**6.23**	**5.29**	**4.77**	**4.44**	**4.20**	**4.03**	**3.89**	**3.78**	**3.69**	**3.55**	**3.41**	**3.26**	**3.18**	**3.10**	**3.02**	**2.93**	**2.75**
	0.0005	0.001	16.1	11.0	9.00	7.94	7.27	6.81	6.46	6.19	5.98	5.81	5.55	5.27	4.99	4.85	4.70	4.54	4.39	4.06
	0.125	0.25	1.42	1.51	1.50	1.49	1.47	1.46	1.45	1.44	1.43	1.43	1.41	1.40	1.39	1.38	1.37	1.36	1.35	1.33
	0.05	0.10	3.03	2.64	2.44	2.31	2.22	2.15	2.10	2.06	2.03	2.00	1.96	1.91	1.86	1.84	1.81	1.78	1.75	1.69
	0.025	**0.05**	**4.45**	**3.59**	**3.20**	**2.96**	**2.81**	**2.70**	**2.61**	**2.55**	**2.49**	**2.45**	**2.38**	**2.31**	**2.23**	**2.19**	**2.15**	**2.10**	**2.06**	**1.96**
17	0.0125	0.025	6.04	4.62	4.01	3.66	3.44	3.28	3.16	3.06	2.98	2.92	2.82	2.72	2.62	2.56	2.50	2.44	2.38	2.25
	0.005	**0.01**	**8.40**	**6.11**	**5.18**	**4.67**	**4.34**	**4.10**	**3.93**	**3.79**	**3.68**	**3.59**	**3.46**	**3.31**	**3.16**	**3.08**	**3.00**	**2.92**	**2.83**	**2.65**
	0.0005	0.001	15.7	10.7	8.73	7.68	7.02	6.56	6.22	5.96	5.75	5.58	5.32	5.05	4.78	4.63	4.48	4.33	4.18	3.85

df for denominator	α_1	α_2	*df for numerator* **1**	**2**	**3**	**4**	**5**	**6**	**7**	**8**	**9**	**10**	**12**	**15**	**20**	**24**	**30**	**40**	**60**	∞
	0.125	0.25	1.41	1.50	1.49	1.48	1.46	1.45	1.44	1.43	1.42	1.42	1.40	1.39	1.38	1.37	1.36	1.35	1.34	1.32
	0.05	0.10	3.01	2.62	2.42	2.29	2.20	2.13	2.08	2.04	2.00	1.98	1.93	1.89	1.84	1.81	1.78	1.75	1.72	1.66
	0.025	**0.05**	**4.41**	**3.55**	**3.16**	**2.93**	**2.77**	**2.66**	**2.58**	**2.51**	**2.46**	**2.41**	**2.34**	**2.27**	**2.19**	**2.15**	**2.11**	**2.06**	**2.02**	**1.92**
18	0.0125	0.025	5.98	4.56	3.95	3.61	3.38	3.22	3.10	3.01	2.93	2.87	2.77	2.67	2.56	2.50	2.44	2.38	2.32	2.19
	0.005	**0.01**	**8.29**	**6.01**	**5.09**	**4.58**	**4.25**	**4.01**	**3.84**	**3.71**	**3.60**	**3.51**	**3.37**	**3.23**	**3.08**	**3.00**	**2.92**	**2.84**	**2.75**	**2.57**
	0.0005	0.001	15.4	10.4	8.49	7.46	6.81	6.35	6.02	5.76	5.56	5.39	5.13	4.87	4.59	4.45	4.30	4.15	4.00	3.67
	0.125	0.25	1.41	1.49	1.49	1.47	1.46	1.44	1.43	1.42	1.41	1.41	1.40	1.38	1.37	1.36	1.35	1.34	1.33	1.30
	0.05	0.10	2.99	2.61	2.40	2.27	2.18	2.11	2.06	2.02	1.98	1.96	1.91	1.86	1.81	1.79	1.76	1.73	1.70	1.63
	0.025	**0.05**	**4.38**	**3.52**	**3.13**	**2.90**	**2.74**	**2.63**	**2.54**	**2.48**	**2.42**	**2.38**	**2.31**	**2.23**	**2.16**	**2.11**	**2.07**	**2.03**	**1.98**	**1.88**
19	0.0125	0.025	5.92	4.51	3.90	3.56	3.33	3.17	3.05	2.96	2.88	2.82	2.72	2.62	2.51	2.45	2.39	2.33	2.27	2.13
	0.005	**0.01**	**8.18**	**5.93**	**5.01**	**4.50**	**4.17**	**3.94**	**3.77**	**3.63**	**3.52**	**3.43**	**3.30**	**3.15**	**3.00**	**2.92**	**2.84**	**2.76**	**2.67**	**2.49**
	0.0005	0.001	15.1	10.2	8.28	7.26	6.62	6.18	5.85	5.59	5.39	5.22	4.97	4.70	4.43	4.29	4.14	3.99	3.84	3.51
	0.125	0.25	1.40	1.49	1.48	1.47	1.45	1.44	1.43	1.42	1.41	1.40	1.39	1.37	1.36	1.35	1.34	1.33	1.32	1.29
	0.05	0.10	2.97	2.59	2.38	2.25	2.16	2.09	2.04	2.00	1.96	1.94	1.89	1.84	1.79	1.77	1.74	1.71	1.68	1.61
	0.025	**0.05**	**4.35**	**3.49**	**3.10**	**2.87**	**2.71**	**2.60**	**2.51**	**2.45**	**2.39**	**2.35**	**2.28**	**2.20**	**2.12**	**2.08**	**2.04**	**1.99**	**1.95**	**1.84**
20	0.0125	0.025	5.87	4.46	3.86	3.51	3.29	3.13	3.01	2.91	2.84	2.77	2.68	2.57	2.46	2.41	2.35	2.29	2.22	2.09
	0.005	**0.01**	**8.10**	**5.85**	**4.94**	**4.43**	**4.10**	**3.87**	**3.70**	**3.56**	**3.46**	**3.37**	**3.23**	**3.09**	**2.94**	**2.86**	**2.78**	**2.69**	**2.61**	**2.42**
	0.0005	0.001	14.8	9.95	8.10	7.10	6.46	6.02	5.69	5.44	5.24	5.08	4.82	4.56	4.29	4.15	4.00	3.86	3.70	3.38
	0.125	0.25	1.40	1.48	1.47	1.45	1.44	1.42	1.41	1.40	1.39	1.39	1.37	1.36	1.34	1.33	1.32	1.31	1.30	1.28
	0.05	0.10	2.95	2.56	2.35	2.22	2.13	2.06	2.01	1.97	1.93	1.90	1.86	1.81	1.76	1.73	1.70	1.67	1.64	1.57
	0.025	**0.05**	**4.30**	**3.44**	**3.05**	**2.82**	**2.66**	**2.55**	**2.46**	**2.40**	**2.34**	**2.30**	**2.23**	**2.15**	**2.07**	**2.03**	**1.98**	**1.94**	**1.89**	**1.78**
22	0.0125	0.025	5.79	4.38	3.78	3.44	3.22	3.05	2.93	2.84	2.76	2.70	2.60	2.50	2.39	2.33	2.27	2.21	2.14	2.00
	0.005	**0.01**	**7.95**	**5.72**	**4.82**	**4.31**	**3.99**	**3.76**	**3.59**	**3.45**	**3.35**	**3.26**	**3.12**	**2.98**	**2.83**	**2.75**	**2.67**	**2.58**	**2.50**	**2.31**
	0.0005	0.001	14.4	9.61	7.80	6.81	6.19	5.76	5.44	5.19	4.99	4.83	4.58	4.33	4.06	3.92	3.78	3.63	3.48	3.15
	0.125	0.25	1.39	1.47	1.46	1.44	1.43	1.41	1.40	1.39	1.38	1.38	1.36	1.35	1.33	1.32	1.31	1.30	1.29	1.26
	0.05	0.10	2.93	2.54	2.33	2.19	2.10	2.04	1.98	1.94	1.91	1.88	1.83	1.78	1.73	1.70	1.67	1.64	1.61	1.53
	0.025	**0.05**	**4.26**	**3.40**	**3.01**	**2.78**	**2.62**	**2.51**	**2.42**	**2.36**	**2.30**	**2.25**	**2.18**	**2.11**	**2.03**	**1.98**	**1.94**	**1.89**	**1.84**	**1.73**
24	0.0125	0.025	5.72	4.32	3.72	3.38	3.15	2.99	2.87	2.78	2.70	2.64	2.54	2.44	2.33	2.27	2.21	2.15	2.08	1.94
	0.005	**0.01**	**7.82**	**5.61**	**4.72**	**4.22**	**3.90**	**3.67**	**3.50**	**3.36**	**3.26**	**3.17**	**3.03**	**2.89**	**2.74**	**2.66**	**2.58**	**2.49**	**2.40**	**2.21**
	0.0005	0.001	14.0	9.34	7.55	6.59	5.98	5.55	5.23	4.99	4.80	4.64	4.39	4.14	3.87	3.74	3.59	3.45	3.29	2.97

df for denominator	α_1	α_2	df for numerator 1	2	3	4	5	6	7	8	9	10	12	15	20	24	30	40	60	∞
	0.125	0.25	1.38	1.46	1.45	1.44	1.42	1.41	1.39	1.38	1.37	1.37	1.35	1.34	1.32	1.31	1.30	1.29	1.28	1.25
	0.05	0.10	2.91	2.52	2.31	2.17	2.08	2.01	1.96	1.92	1.88	1.86	1.81	1.76	1.71	1.68	1.65	1.61	1.58	1.50
	0.025	**0.05**	**4.23**	**3.37**	**2.98**	**2.74**	**2.59**	**2.47**	**2.39**	**2.32**	**2.27**	**2.22**	**2.15**	**2.07**	**1.99**	**1.95**	**1.90**	**1.85**	**1.80**	**1.69**
26	0.0125	0.025	5.66	4.27	3.67	3.33	3.10	2.94	2.82	2.73	2.65	2.59	2.49	2.39	2.28	2.22	2.16	2.09	2.03	1.88
	0.005	**0.01**	**7.72**	**5.53**	**4.64**	**4.14**	**3.82**	**3.59**	**3.42**	**3.29**	**3.18**	**3.09**	**2.96**	**2.81**	**2.66**	**2.58**	**2.50**	**2.42**	**2.33**	**2.13**
	0.0005	0.001	13.7	9.12	7.36	6.41	5.80	5.38	5.07	4.83	4.64	4.48	4.24	3.99	3.72	3.59	3.44	3.30	3.15	2.82
	0.125	0.25	1.38	1.46	1.45	1.43	1.41	1.40	1.39	1.38	1.37	1.36	1.34	1.33	1.31	1.30	1.29	1.28	1.27	1.24
	0.05	0.10	2.89	2.50	2.29	2.16	2.06	2.00	1.94	1.90	1.87	1.84	1.79	1.74	1.69	1.66	1.63	1.59	1.56	1.48
	0.025	**0.05**	**4.20**	**3.34**	**2.95**	**2.71**	**2.56**	**2.45**	**2.36**	**2.29**	**2.24**	**2.19**	**2.12**	**2.04**	**1.96**	**1.91**	**1.87**	**1.82**	**1.77**	**1.65**
28	0.0125	0.025	5.61	4.22	3.63	3.29	3.06	2.90	2.78	2.69	2.61	2.55	2.45	2.34	2.23	2.17	2.11	2.05	1.98	1.83
	0.005	**0.01**	**7.64**	**5.45**	**4.57**	**4.07**	**3.75**	**3.53**	**3.36**	**3.23**	**3.12**	**3.03**	**2.90**	**2.75**	**2.60**	**2.52**	**2.44**	**2.35**	**2.26**	**2.06**
	0.0005	0.001	13.5	8.93	7.19	6.25	5.66	5.24	4.93	4.69	4.50	4.35	4.11	3.86	3.60	3.46	3.32	3.18	3.02	2.69
	0.125	0.25	1.38	1.45	1.44	1.42	1.41	1.39	1.38	1.37	1.36	1.35	1.34	1.32	1.30	1.29	1.28	1.27	1.26	1.23
	0.05	0.10	2.88	2.49	2.28	2.14	2.05	1.98	1.93	1.88	1.85	1.82	1.77	1.72	1.67	1.64	1.61	1.57	1.54	1.46
	0.025	**0.05**	**4.17**	**3.32**	**2.92**	**2.69**	**2.53**	**2.42**	**2.33**	**2.27**	**2.21**	**2.16**	**2.09**	**2.01**	**1.93**	**1.89**	**1.84**	**1.79**	**1.74**	**1.62**
30	0.0125	0.025	5.57	4.18	3.59	3.25	3.03	2.87	2.75	2.65	2.57	2.51	2.41	2.31	2.20	2.14	2.07	2.01	1.94	1.79
	0.005	**0.01**	**7.56**	**5.39**	**4.51**	**4.02**	**3.70**	**3.47**	**3.30**	**3.17**	**3.07**	**2.98**	**2.84**	**2.70**	**2.55**	**2.47**	**2.39**	**2.30**	**2.21**	**2.01**
	0.0005	0.001	13.3	8.77	7.05	6.12	5.53	5.12	4.82	4.58	4.39	4.24	4.00	3.75	3.49	3.36	3.22	3.07	2.92	2.59
	0.125	0.25	1.36	1.44	1.42	1.40	1.39	1.37	1.36	1.35	1.34	1.33	1.31	1.30	1.28	1.26	1.25	1.24	1.22	1.19
	0.05	0.10	2.84	2.44	2.23	2.09	2.00	1.93	1.87	1.83	1.79	1.76	1.71	1.66	1.61	1.57	1.54	1.51	1.47	1.38
	0.025	**0.05**	**4.08**	**3.23**	**2.84**	**2.61**	**2.45**	**2.34**	**2.25**	**2.18**	**2.12**	**2.08**	**2.00**	**1.92**	**1.84**	**1.79**	**1.74**	**1.69**	**1.64**	**1.51**
40	0.0125	0.025	5.42	4.05	3.46	3.13	2.90	2.74	2.62	2.53	2.45	2.39	2.29	2.18	2.07	2.01	1.94	1.88	1.80	1.64
	0.005	**0.01**	**7.31**	**5.18**	**4.31**	**3.83**	**3.51**	**3.29**	**3.12**	**2.99**	**2.89**	**2.80**	**2.66**	**2.52**	**2.37**	**2.29**	**2.20**	**2.11**	**2.02**	**1.80**
	0.0005	0.001	12.6	8.25	6.60	5.70	5.13	4.73	4.44	4.21	4.02	3.87	3.64	3.40	3.15	3.01	2.87	2.73	2.57	2.23

df for denominator	α_1	α_2	*df for numerator* 1	2	3	4	5	6	7	8	9	10	12	15	20	24	30	40	60	∞
	0.125	0.25	1.35	1.42	1.41	1.38	1.37	1.35	1.33	1.32	1.31	1.30	1.29	1.27	1.25	1.24	1.22	1.21	1.19	1.15
	0.05	0.10	2.79	2.39	2.18	2.04	1.95	1.87	1.82	1.77	1.74	1.71	1.66	1.60	1.54	1.51	1.48	1.44	1.40	1.29
	0.025	**0.05**	**4.00**	**3.15**	**2.76**	**2.53**	**2.37**	**2.25**	**2.17**	**2.10**	**2.04**	**1.99**	**1.92**	**1.84**	**1.75**	**1.70**	**1.65**	**1.59**	**1.53**	**1.39**
60	0.0125	0.025	5.29	3.93	3.34	3.01	2.79	2.63	2.51	2.41	2.33	2.27	2.17	2.06	1.94	1.88	1.82	1.74	1.67	1.48
	0.005	**0.01**	**7.08**	**4.98**	**4.13**	**3.65**	**3.34**	**3.12**	**2.95**	**2.82**	**2.72**	**2.63**	**2.50**	**2.35**	**2.20**	**2.12**	**2.03**	**1.94**	**1.84**	**1.60**
	0.0005	0.001	12.0	7.76	6.17	5.31	4.76	4.37	4.09	3.87	3.69	3.54	3.31	3.08	2.83	2.69	2.55	2.41	2.25	1.89
	0.125	0.25	1.34	1.40	1.39	1.37	1.35	1.33	1.31	1.30	1.29	1.28	1.26	1.24	1.22	1.21	1.19	1.18	1.16	1.10
	0.05	0.10	2.75	2.35	2.13	1.99	1.90	1.82	1.77	1.72	1.68	1.65	1.60	1.55	1.48	1.45	1.41	1.37	1.32	1.19
	0.025	**0.05**	**3.92**	**3.07**	**2.68**	**2.45**	**2.29**	**2.17**	**2.09**	**2.02**	**1.96**	**1.91**	**1.83**	**1.75**	**1.66**	**1.61**	**1.55**	**1.50**	**1.43**	**1.25**
120	0.0125	0.025	5.15	3.80	3.23	2.89	2.67	2.52	2.39	2.30	2.22	2.16	2.05	1.94	1.82	1.76	1.69	1.61	1.53	1.31
	0.005	**0.01**	**6.85**	**4.79**	**3.95**	**3.48**	**3.17**	**2.96**	**2.79**	**2.66**	**2.56**	**2.47**	**2.34**	**2.19**	**2.03**	**1.95**	**1.86**	**1.76**	**1.66**	**1.38**
	0.0005	0.001	11.4	7.32	5.79	4.95	4.42	4.04	3.77	3.55	3.38	3.24	3.02	2.78	2.53	2.40	2.26	2.11	1.95	1.54
	0.125	0.25	1.32	1.39	1.37	1.35	1.33	1.31	1.29	1.28	1.27	1.25	1.24	1.22	1.19	1.18	1.16	1.14	1.12	1.00
	0.05	0.10	2.71	2.30	2.08	1.94	1.85	1.77	1.72	1.67	1.63	1.60	1.55	1.49	1.42	1.38	1.34	1.30	1.24	1.00
	0.025	**0.05**	**3.84**	**3.00**	**2.60**	**2.37**	**2.21**	**2.10**	**2.01**	**1.94**	**1.88**	**1.83**	**1.75**	**1.67**	**1.57**	**1.52**	**1.46**	**1.39**	**1.32**	**1.00**
∞	0.0125	0.025	5.02	3.69	3.12	2.79	2.57	2.41	2.29	2.19	2.11	2.05	1.94	1.83	1.71	1.64	1.57	1.48	1.39	1.00
	0.005	**0.01**	**6.63**	**4.61**	**3.78**	**3.32**	**3.02**	**2.80**	**2.64**	**2.51**	**2.41**	**2.32**	**2.18**	**2.04**	**1.88**	**1.79**	**1.70**	**1.59**	**1.47**	**1.00**
	0.0005	0.001	10.8	6.91	5.42	4.62	4.10	3.74	3.47	3.27	3.10	2.96	2.74	2.51	2.27	2.13	1.99	1.84	1.66	1.00

Table F.3 Non-central F distribution

		Power = 1 − (tabled entry)										
							ϕ					
df_2	α	0.50	1.0	1.2	1.4	1.6	1.8	2.0	2.2	2.6	3.0	4.0
							$df_1 = 1$					
2	0.05	0.93	0.86	0.83	0.78	0.74	0.69	0.64	0.59	0.49	0.40	0.20
	0.01	0.01	0.97	0.96	0.95	0.94	0.93	0.91	0.90	0.87	0.83	0.72
4	0.05	0.91	0.80	0.74	0.67	0.59	0.51	0.43	0.35	0.22	0.12	0.02
	0.01	0.98	0.95	0.94	0.92	0.89	0.86	0.82	0.78	0.67	0.56	0.23
6	0.05	0.91	0.78	0.70	0.62	0.52	0.43	0.34	0.26	0.14	0.06	0.00
	0.01	0.98	0.93	0.90	0.86	0.81	0.75	0.69	0.61	0.46	0.31	0.08
8	0.05	0.90	0.76	0.68	0.59	0.49	0.39	0.30	0.22	0.11	0.04	0.00
	0.01	0.98	0.92	0.89	0.84	0.78	0.70	0.62	0.54	0.37	0.22	0.03
10	0.05	0.90	0.75	0.66	0.57	0.47	0.37	0.28	0.20	0.09	0.03	0.00
	0.01	0.98	0.92	0.87	0.82	0.75	0.67	0.58	0.49	0.31	0.17	0.02
12	0.05	0.90	0.74	0.65	0.56	0.45	0.35	0.26	0.19	0.08	0.03	0.00
	0.01	0.97	0.91	0.87	0.81	0.73	0.65	0.55	0.46	0.28	0.14	0.00
16	0.05	0.90	0.74	0.64	0.54	0.43	0.33	0.24	0.17	0.07	0.02	0.00
	0.01	0.97	0.90	0.85	0.79	0.71	0.61	0.52	0.42	0.24	0.11	0.00
20	0.05	0.90	0.73	0.63	0.53	0.42	0.32	0.23	0.16	0.06	0.02	0.00
	0.01	0.97	0.90	0.85	0.78	0.69	0.59	0.49	0.39	0.21	0.10	0.00
30	0.05	0.89	0.72	0.62	0.52	0.40	0.31	0.22	0.15	0.06	0.02	0.00
	0.01	0.97	0.89	0.83	0.76	0.67	0.57	0.46	0.36	0.19	0.08	0.00
∞	0.05	0.89	0.71	0.70	0.49	0.38	0.28	0.19	0.12	0.04	0.01	0.00
	0.01	0.97	0.88	0.81	0.72	0.62	0.51	0.40	0.30	0.14	0.05	0.00
df_2	α						$df_1 = 2$					
2	0.05	0.93	0.88	0.85	0.82	0.78	0.75	0.70	0.66	0.57	0.48	0.29
	0.01	0.99	0.98	0.97	0.96	0.95	0.94	0.93	0.92	0.89	0.86	0.78
4	0.05	0.92	0.82	0.77	0.70	0.62	0.54	0.46	0.38	0.24	0.14	0.02
	0.01	0.98	0.96	0.94	0.92	0.89	0.85	0.81	0.76	0.66	0.54	0.27
6	0.05	0.91	0.79	0.71	0.63	0.53	0.43	0.34	0.26	0.13	0.05	0.00
	0.01	0.98	0.94	0.91	0.87	0.82	0.76	0.70	0.62	0.46	0.31	0.07
8	0.05	0.91	0.77	0.68	0.58	0.48	0.37	0.28	0.20	0.08	0.03	0.00
	0.01	0.98	0.93	0.89	0.84	0.78	0.70	0.61	0.52	0.34	0.19	0.02
10	0.05	0.91	0.75	0.66	0.55	0.44	0.34	0.24	0.16	0.06	0.02	0.00
	0.01	0.98	0.92	0.88	0.82	0.74	0.65	0.55	0.45	0.26	0.13	0.01
12	0.05	0.90	0.74	0.64	0.53	0.42	0.31	0.22	0.14	0.05	0.01	0.00
	0.01	0.98	0.91	0.86	0.80	0.71	0.61	0.51	0.40	0.22	0.09	0.00
16	0.05	0.90	0.73	0.62	0.51	0.39	0.28	0.19	0.12	0.04	0.01	0.00
	0.01	0.97	0.90	0.84	0.77	0.67	0.57	0.45	0.34	0.16	0.06	0.00
20	0.05	0.90	0.72	0.61	0.49	0.37	0.26	0.17	0.11	0.03	0.01	0.00
	0.01	0.97	0.90	0.83	0.75	0.65	0.53	0.42	0.31	0.14	0.04	0.00
30	0.05	0.90	0.71	0.59	0.47	0.35	0.24	0.15	0.09	0.02	0.00	0.00
	0.01	0.97	0.88	0.82	0.72	0.61	0.49	0.37	0.26	0.10	0.03	0.00
∞	0.05	0.89	0.68	0.56	0.43	0.30	0.20	0.12	0.06	0.01	0.00	0.00
	0.01	0.97	0.86	0.77	0.66	0.53	0.40	0.28	0.18	0.05	0.01	0.00

df_1 = degrees of freedom between groups
df_2 = degrees of freedom within groups

Table F.3 Non-central F distribution *contd*

		Power = 1 − (tabled entry)										
		ϕ										
df_2	α	0.50	1.0	1.2	1.4	1.6	1.8	2.0	2.2	2.6	3.0	4.0
		$df_1 = 3$										
2	0.05	0.93	0.89	0.86	0.83	0.80	0.76	0.73	0.69	0.60	0.52	0.32
	0.01	0.99	0.98	0.97	0.96	0.96	0.95	0.94	0.93	0.90	0.88	0.80
4	0.05	0.92	0.83	0.77	0.71	0.63	0.55	0.47	0.39	0.25	0.14	0.02
	0.01	0.98	0.96	0.94	0.92	0.89	0.86	0.82	0.77	0.67	0.55	0.28
6	0.05	0.91	0.79	0.71	0.62	0.52	0.42	0.33	0.24	0.11	0.04	0.00
	0.01	0.98	0.94	0.91	0.87	0.82	0.76	0.69	0.61	0.44	0.29	0.06
8	0.05	0.91	0.76	0.67	0.57	0.46	0.35	0.25	0.17	0.06	0.02	0.00
	0.01	0.98	0.93	0.89	0.84	0.77	0.68	0.59	0.49	0.30	0.16	0.01
10	0.05	0.91	0.75	0.65	0.53	0.41	0.30	0.21	0.13	0.04	0.01	0.00
	0.01	0.98	0.92	0.87	0.80	0.72	0.62	0.52	0.41	0.22	0.09	0.00
12	0.05	0.90	0.73	0.62	0.51	0.38	0.27	0.18	0.11	0.03	0.01	0.00
	0.01	0.98	0.91	0.85	0.78	0.69	0.58	0.46	0.35	0.17	0.06	0.00
16	0.05	0.90	0.71	0.60	0.47	0.34	0.23	0.14	0.08	0.02	0.00	0.00
	0.01	0.97	0.90	0.83	0.74	0.64	0.51	0.39	0.28	0.11	0.03	0.00
20	0.05	0.90	0.70	0.58	0.45	0.32	0.21	0.13	0.07	0.01	0.00	0.00
	0.01	0.97	0.89	0.82	0.72	0.60	0.47	0.35	0.24	0.08	0.02	0.00
30	0.05	0.89	0.68	0.55	0.42	0.29	0.18	0.10	0.05	0.01	0.00	0.00
	0.01	0.97	0.87	0.79	0.68	0.55	0.42	0.29	0.18	0.05	0.01	0.00
∞	0.05	0.88	0.64	0.50	0.36	0.23	0.13	0.07	0.03	0.00	0.00	0.00
	0.01	0.97	0.84	0.73	0.59	0.44	0.30	0.18	0.10	0.02	0.00	0.00
df_2	α	$df_1 = 4$										
2	0.05	0.94	0.89	0.87	0.84	0.81	0.77	0.74	0.70	0.62	0.54	0.34
	0.01	0.99	0.98	0.97	0.97	0.96	0.95	0.94	0.93	0.91	0.88	0.81
4	0.05	0.92	0.83	0.78	0.71	0.64	0.55	0.47	0.39	0.25	0.14	0.02
	0.01	0.98	0.96	0.94	0.92	0.89	0.86	0.82	0.78	0.67	0.56	0.28
6	0.05	0.92	0.79	0.71	0.62	0.52	0.41	0.31	0.23	0.10	0.04	0.00
	0.01	0.98	0.94	0.91	0.87	0.82	0.76	0.68	0.60	0.43	0.28	0.05
8	0.05	0.91	0.76	0.66	0.55	0.44	0.33	0.23	0.15	0.05	0.01	0.00
	0.01	0.98	0.93	0.89	0.83	0.76	0.67	0.57	0.47	0.28	0.14	0.01
10	0.05	0.91	0.74	0.63	0.51	0.39	0.27	0.18	0.11	0.03	0.01	0.00
	0.01	0.98	0.92	0.87	0.79	0.70	0.60	0.49	0.37	0.19	0.07	0.00
12	0.05	0.90	0.72	0.61	0.48	0.35	0.24	0.15	0.08	0.02	0.00	0.00
	0.01	0.98	0.91	0.85	0.76	0.66	0.55	0.42	0.31	0.13	0.04	0.00
16	0.05	0.90	0.70	0.57	0.44	0.31	0.19	0.11	0.06	0.01	0.00	0.00
	0.01	0.97	0.89	0.82	0.72	0.60	0.47	0.34	0.23	0.08	0.02	0.00
20	0.05	0.89	0.68	0.55	0.41	0.28	0.17	0.09	0.04	0.01	0.00	0.00
	0.01	0.97	0.88	0.80	0.69	0.56	0.42	0.29	0.18	0.05	0.01	0.00
30	0.05	0.89	0.66	0.52	0.37	0.24	0.14	0.07	0.03	0.00	0.00	0.00
	0.01	0.97	0.86	0.77	0.64	0.50	0.35	0.22	0.13	0.03	0.00	0.00
∞	0.05	0.88	0.60	0.45	0.29	0.17	0.08	0.04	0.01	0.00	0.00	0.00
	0.01	0.96	0.81	0.68	0.53	0.36	0.22	0.11	0.05	0.01	0.00	0.00

Table F.4 Distribution of the Studentized range statistic

df	α	$r = 1 +$ *number of steps between ordered means*													
		2	*3*	*4*	*5*	*6*	*7*	*8*	*9*	*10*	*11*	*12*	*13*	*14*	*15*
1	0.05	18.0	27.0	32.8	37.1	40.4	43.1	45.4	47.4	49.1	50.6	52.0	53.2	54.3	55.4
	0.01	90.0	135	164	186	202	216	227	237	246	253	260	266	272	277
2	0.05	6.09	8.3	9.8	10.9	11.7	12.4	13.0	13.5	14.0	14.4	14.7	15.1	15.4	15.7
	0.01	14.0	19.0	22.3	24.7	26.6	28.2	29.5	30.7	31.7	32.6	33.4	34.1	34.8	35.4
3	0.05	4.50	5.91	6.82	7.50	8.04	8.48	8.85	9.18	9.46	9.72	9.95	10.2	10.4	10.5
	0.01	8.26	10.6	12.2	13.3	14.2	15.0	15.6	16.2	16.7	17.1	17.5	17.9	18.2	18.5
4	0.05	3.93	5.04	5.76	6.29	6.71	7.05	7.35	7.60	7.83	8.03	8.21	8.37	8.52	8.66
	0.01	6.51	8.12	9.17	9.96	10.6	11.1	11.5	11.9	12.3	12.6	12.8	13.1	13.3	13.5
5	0.05	3.64	4.60	5.22	5.67	6.03	6.33	6.58	6.80	6.99	7.17	7.32	7.47	7.60	7.72
	0.01	5.70	6.97	7.80	8.42	8.91	9.32	9.67	9.97	10.2	10.5	10.7	10.9	11.1	11.2
6	0.05	3.46	4.34	4.90	5.31	5.63	5.89	6.12	6.32	6.49	6.65	6.79	6.92	7.03	7.14
	0.01	5.24	6.33	7.03	7.56	7.97	8.32	8.61	8.87	9.10	9.30	9.49	9.65	9.81	9.95
7	0.05	3.34	4.16	4.69	5.06	5.36	5.61	5.82	6.00	6.16	6.30	6.43	6.55	6.66	6.76
	0.01	4.95	5.92	6.54	7.01	7.37	7.68	7.94	8.17	8.37	8.55	8.71	8.86	9.00	9.12
8	0.05	3.26	4.04	4.53	4.89	5.17	5.40	5.60	5.77	5.92	6.05	6.18	6.29	6.39	6.48
	0.01	4.74	5.63	6.20	6.63	6.96	7.24	7.47	7.68	7.87	8.03	8.18	8.31	8.44	8.55
9	0.05	3.20	3.95	4.42	4.76	5.02	5.24	5.43	5.60	5.74	5.87	5.98	6.09	6.19	6.28
	0.01	4.60	5.43	5.96	6.35	6.66	6.91	7.13	7.32	7.49	7.65	7.78	7.91	8.03	8.13
10	0.05	3.15	3.88	4.33	4.65	4.91	5.12	5.30	5.46	5.60	5.72	5.83	5.93	6.03	6.11
	0.01	4.48	5.27	5.77	6.14	6.43	6.67	6.87	7.05	7.21	7.36	7.48	7.60	7.71	7.81
11	0.05	3.11	3.82	4.26	4.57	4.82	5.03	5.20	5.35	5.49	5.61	5.71	5.81	5.90	5.99
	0.01	4.39	5.14	5.62	5.97	6.25	6.48	6.67	6.84	6.99	7.13	7.26	7.36	7.46	7.56
12	0.05	3.08	3.77	4.20	4.51	4.75	4.95	5.12	5.27	5.40	5.51	5.62	5.71	5.80	5.88
	0.01	4.32	5.04	5.50	5.84	6.10	6.32	6.51	6.67	6.81	6.94	7.06	7.17	7.26	7.36

Table F.4 Distribution of the Studentized range statistic *contd*

df	α	*r = 1 + number of steps between ordered means*													
		2	*3*	*4*	*5*	*6*	*7*	*8*	*9*	*10*	*11*	*12*	*13*	*14*	*15*
13	0.05	3.06	3.73	4.15	4.45	4.69	4.88	5.05	5.19	5.32	5.43	5.53	5.63	5.71	5.79
	0.01	4.26	4.96	5.40	5.73	5.98	6.19	6.37	6.53	6.67	6.79	6.90	7.01	7.10	7.19
14	0.05	3.03	3.70	4.11	4.41	4.64	4.83	4.99	5.13	5.25	5.36	5.46	5.55	6.64	5.72
	0.01	4.21	4.89	5.32	5.63	5.88	6.08	6.26	6.41	6.54	6.66	6.77	6.87	6.96	7.05
16	0.05	3.00	3.65	4.05	4.33	4.56	4.74	4.90	5.03	5.15	5.26	5.35	5.44	5.52	5.59
	0.01	4.13	4.78	5.19	5.49	5.72	5.92	6.08	6.22	6.35	6.46	6.56	6.66	6.74	6.82
18	0.05	2.97	3.61	4.00	4.28	4.49	4.67	4.82	4.96	5.07	5.17	5.27	5.35	5.43	5.50
	0.01	4.07	4.70	5.09	5.38	5.60	5.79	5.94	6.08	6.20	6.31	6.41	6.50	6.58	6.65
20	0.05	2.95	3.58	3.96	4.23	4.45	4.62	4.77	4.90	5.01	5.11	5.20	5.28	5.36	5.43
	0.01	4.02	4.64	5.02	5.29	5.51	5.69	5.84	5.97	6.09	6.19	6.29	6.37	6.45	6.52
24	0.05	2.92	3.53	3.90	4.17	4.37	4.54	4.68	4.81	4.92	5.01	5.10	5.18	5.25	5.32
	0.01	3.96	4.54	4.91	5.17	5.37	5.54	5.69	5.81	5.92	6.02	6.11	6.19	6.26	6.33
30	0.05	2.89	3.49	3.84	4.10	4.30	4.46	4.60	4.72	4.83	4.92	5.00	5.08	5.15	5.21
	0.01	3.89	4.45	4.80	5.05	5.24	5.40	5.54	5.56	5.76	5.85	5.93	6.01	6.08	6.14
40	0.05	2.86	3.44	3.79	4.04	4.23	4.39	4.52	4.63	4.74	4.82	4.91	4.98	5.05	5.11
	0.01	3.82	4.37	4.70	4.93	5.11	5.27	5.39	5.50	5.60	5.69	5.77	5.84	5.90	5.96
60	0.05	2.83	3.40	3.74	3.98	4.16	4.31	4.44	4.55	4.65	4.73	4.81	4.88	4.94	5.00
	0.01	3.76	4.28	4.60	4.82	4.99	5.13	5.25	5.36	5.45	5.53	5.60	5.67	5.73	5.79
120	0.05	2.80	3.36	3.69	3.92	4.10	4.24	4.36	4.48	4.56	4.64	4.72	4.78	4.84	4.90
	0.01	3.70	4.20	4.50	4.71	4.87	5.01	5.12	5.21	5.30	5.38	5.44	5.51	5.56	5.61
∞	0.05	2.77	3.31	3.63	3.86	4.03	4.17	4.29	4.39	4.47	4.55	4.62	4.68	4.74	4.80
	0.01	3.64	4.12	4.40	4.60	4.76	4.88	4.99	5.08	5.16	5.23	5.29	5.35	5.40	5.45

References

Armitage, P. (1987) *Statistical Methods in Medical Research*, 2nd edn, Blackwell, Oxford.
Healey, M.J.R. (1988) *GLIM: An Introduction*, Clarendon Press, Oxford.
Knoke, D. and Burke, P.J. (1983) *Log-linear Models*, Sage, London.
Maxwell, S.E. and Delaney, H.D. (1990) *Designing Experiments and Analysing Data*, Wadsworth, California.
Postman, L. and Keppel, G. (1977) *Journal of Experimental Psychology*.
Satterthwaite, F.E. (1946) An approximate distribution of estimates of variance components, *Biometrics Bulletin*, **2**, 110–14.
Scheffé, H.A. (1959) *The Analysis of Variance*, Wiley, New York.
Shore, M.F. (1958) Perceptual efficiency as related to induced muscular effort and manifest anxiety, *Journal of Experimental Psychology*, **55**, 179–83.
Winer, B.J., Brown, D.R. and Michels, K.M. (1991) *Statistical Principles in Experiment Design*, 3rd edn, McGraw-Hill, New York.

Index